CHEMICAL MAGIC

SECOND EDITION

Leonard A. Ford

Revised by
E. Winston Grundmeier

DOVER PUBLICATIONS, INC.
New York

WARNING

IMPORTANT SAFETY NOTICE

Demonstrations and experiments in this book marked with a bullet (•) should be performed only by those trained sufficiently well in chemistry to know exactly how to handle the potentially toxic, corrosive or explosive substances used therein; minors and novices should not attempt these without strict knowledgeable supervision.

Users should also be alert to the possibility that some substances mentioned in this book, previously thought harmless, may be discovered to possess harmful properties.

Bibliographical Note

This Dover edition, first published in 1993, is a revised republication of the work originally written by Leonard A. Ford and published by T. S. Denison & Company, Minneapolis, 1959. For this edition, E. Winston Grundmeier has revised the introductory material and many of the demonstrations, deleted some demonstrations and added others. He has also added a new Preface.

Library of Congress Cataloging-in-Publication Data

Ford, Leonard Augustine, 1904–
    Chemical magic / Leonard A. Ford. — 2nd ed. / rev. by E. Winston Grundmeier.
        p.   cm.
    ISBN 0-486-67628-5
    1. Chemistry—Experiments. I. Grundmeier, E. Winston. II. Title.
QD43.F6 1993
542—dc20   92-42557
                                                        CIP

Manufactured in the United States of America
Dover Publications, Inc., 31 East 2nd Street, Mineola, N.Y. 11501

Preface to the Dover Edition

When this book first came out in 1959, it soon began to enjoy widespread popularity as a manual for "chemical magicians," at both the secondary-school and college levels. The book was also printed in at least two foreign languages, and so attained a degree of international recognition. For many years its popular-level, easy-to-understand descriptions of chemical stunts and demonstrations made it one of the most widely used manuals in its field. Each demonstration or "trick," moreover, was genuine science, not mere deception, and always included an explanation of the principles involved. It is the purpose of this edition to bring back Dr. Ford's classic text in revised and updated form.

The present author (or, primarily, reviser) has had the benefit of many years' acquaintance with Dr. Ford and his Chemical Magic shows. As a student at Mankato State Teachers' College (now Mankato State University), I participated as a demonstrator in the first few of Dr. Ford's shows, which he had originated at Mankato State. Later, when I returned to teach there, I gained a greater appreciation of the enormous amount of time and study that Dr. Ford put into his "traveling shows for science," as he called his demonstrations before service clubs, youth organization, church groups, schools, etc. Dr. Ford firmly believed that the interest of the future scientist in science might first be sparked by the "magic" of these shows. I know one case where he was definitely right: it happened to me!

The present revision is based on Dr. Ford's own work. Some replacements of the materials in the demonstrations have been made, largely with an eye to safety—of both the demonstrator and the audience. I have also replaced some entire demonstrations with others, but these replacements come mostly from Dr. Ford's other papers on chemical magic.

For this edition I have grouped the demonstrations according to type, e.g., "color change," "explosions," etc. This should aid the potential user in suiting his or her needs. I have also updated some of the information. I have found it necessary as well to make some substitutions of chemicals to meet modern safety requirements. Overall, the emphasis in this edition has been on greater safety.

—E. WINSTON GRUNDMEIER

Preface to the First Edition

The "Chemical Magic" show is a principal feature of the annual science fair at Mankato State College. Performed by college students, the demonstrations have created interest among science teachers and students. Many requests for information on the procedures have been received. I have explained a number of unusual chemistry demonstrations in the *Proceedings of the Minnesota Academy of Science*, bulletins published by the college, and references to these experiments have appeared in *School Science and Mathematics* and the *Journal of Chemical Education*. It is the large number of requests for detailed information on procedures that has prompted me to explain them in this book. These requests have come from college and high school teachers as well as sponsors of youth organizations who are interested in "chemical magic."

Credit for originating a specific experiment has been impossible to establish. Certainly, experiments similar to those described in the book have been performed countless times by teachers and students in the past. To my knowledge, unusual and spectacular chemical experiments have not been assembled in a book. It is my hope that the experiments described herein will serve as a stimulus for young scientists today.

Every demonstration described has been tested by me and my students. I have attempted to specify procedures and quantities of chemicals that are workable. However, every demonstrator should not hesitate to do some experimentation. It is impossible to make the description of an experiment absolutely foolproof. Therefore, do not give up if the experiment does not work the first time. Spend some time in studying the experiment and reactions involved. You may wish to use more chemicals or less; possibly increase or decrease the size of beakers or flasks.

The possibility of accidents from burns or broken glass has

caused me some concern. I have seen a college chemistry instructor severely burned by white phosphorus. In fifteen years of experimentation with chemical magic type of experiments I have never seen a student injured or harmed. You may eliminate dangerous experiments from a chemical magic show, but explosions and fires produce the most spectacular demonstrations and should be included if the adult in charge of the experiments is willing to set down some hard and fast rules for his student demonstrators. Suggested rules are:

1. Students are permitted to try dangerous experiments only under competent adult supervision.
2. White (or yellow) phosphorus and its solutions must be handled only by a teacher or trained scientist.
3. Public demonstrations will be permitted only after repeated rehearsals.
4. No student is permitted to work alone in a laboratory at any time.

—LEONARD A. FORD

Contents

Introduction

This book contains a collection of experiments or demonstrations magical and mystifying in nature. They can be performed by anyone interested in science. The apparatus and chemicals are found in almost any school laboratory or can be purchased readily from a supply house. Some of the materials can be brought from home. You are given here a group of simple, mystifying demonstrations that can be done with ordinary materials.

Experiments similar to those described have appeared from time to time in the *Journal of Chemical Education, School Science and Mathematics,* in some laboratory manuals and other publications intended for entertainment with science tricks.

All demonstrations have been tested on many groups of adults and students to their amazement and amusement. They have been used at service clubs, youth organizations, church groups, and at science fairs. Performers have been college and high school students and teachers in colleges and high schools. The experiments have been found to be entertaining and informative at assembly programs and in the classroom. These demonstrations have helped to inspire young people who have since made chemistry their life work.

A kit containing chemicals, solutions, beakers and other materials can be taken to group meetings. An interesting half-hour program can be performed with a surprisingly small amount of material by careful selection of demonstrations. The materials for such a program can be put into one or two cardboard boxes. Water to make up solutions can be added when the demonstrator has arrived at the destination.

Mystery Demonstrations

The presentation of the mystery type of experiment is much more important than the experiment itself. The entertainment value lies entirely in the manner of performing, the patter or story that you tell. The audience enjoys being mystified and takes pleasure in contemplating how the demonstration is done.

The difference between a mystery performance for entertainment and a student classroom demonstration is that in the former you intend to have the demonstration remain a mystery after it is shown, whereas in the classroom, the purpose is to seek out a solution.

For entertainment purposes you misdirect the audience. Make false suggestions to deceive the eye as well as the mind. Never make outright falsehoods, but do your experiment in such a way that the audience is misdirected. This will result in amusement, amazement and surprise. But to misdirect, one must practice the experiment until the demonstrator is sure it will work. To do this

1. Study the directions.
2. Assemble all materials.
3. Do the experiment.
4. Make up a story or patter.
5. Rehearse at least once—or more.

Keep your audience in good humor and the true solution to your demonstration will go undetected. This is much more difficult to do with younger students than with adults and scientists, who are often the easiest to mystify.

This book can give only the presentation. More important is showmanship, patter or the art of presentation. That makes the experiment a mystery demonstration. You can acquire the secret only by continual practice.

When used in the classroom or science club, magic- or mystery-type experiments create problems for solution in the mind of the student. Properly presented, the demonstration can serve as a means of solving problems by the use of well-established scientific principles. Often a teacher may wish to perform a classroom experiment without any explanation whatever and leave it to the students to attempt an explanation by experimental means or study. Likewise, a student may use the silent demonstration at the science club to stimulate thought, arouse curiosity and interest.

Be Careful

Flames, explosions and smoke hold the attention of an audience and for that reason are an important part of a group of unusual experiments. Sheets of metal, which can be purchased from a supply house, should be placed under combustible materials to prevent damage to the table top. A carbon dioxide fire extinguisher should be near at hand whenever there is a possibility of flames getting out of control. Solutions of phosphorus in carbon disulfide are difficult to handle since a drop of this liquid will readily cause combustible materials to start burning. Keep a small bottle of this highly combustible substance inside of a glass-stoppered wide-mouth bottle.

Carry out experiments that involve flames or explosions far enough away from the audience so that they are in no danger from noxious gases, burning materials or caustic chemicals.

Broken glass causes bad wounds. When placing a glass tube or glass rod into the hole of a rubber stopper, lubricate first with a drop of glycerine.

Any experiment can be dangerous if the experimenter is clumsy, does not think about what he or she is doing or does not use ordinary caution. The person who breaks things at home will likely do the same when at work with chemicals or glassware. Experience with caustic chemicals, volatile liquids and flammable materials lessens the danger of accidents. Students and teachers alike are urged to practice each experiment before a public presentation to gain confidence and likewise to avoid accidents.

At the end of your performance, inquisitive students may crowd around the demonstration table. They may want to handle equipment, materials and solutions. You must be alert. Keep students away from the demonstration table to avoid accidents. They may spill liquids or ruin your equipment.

Motion, Sound and Color

Movement of a gas, liquid or solid attracts attention. The motion involved in experiments in chemistry should be emphasized. The ammonia fountain experiment becomes spectacular when the upper flask is four or five feet above the lower flask since the motion of the liquid upward is so pronounced. A tall cylinder of bubbling carbon dioxide in a colored solution creates marked attention whereas the gas bubbling in a beaker may not be observed.

The violent explosion that takes place when a potassium chlorate capsule explodes or the hissing sound produced when the ammonia fountain is operating will make these experiments long remembered. The odors of chemicals, the colors of gases, liquids or solids and solutions as well as the movement of particles make demonstrations involving these substances especially useful as a teaching aid or an entertainment feature.

Color liquids with food coloring. Make your mystery demonstrations highly decorative with all the colors of the rainbow. Buy a package of food coloring at the grocery store. The coloring material is not poisonous if you should accidentally swallow colored water. It is not caustic on the skin and will wash off with soap and water.

A typical package of food coloring contains four small bottles. Each bottle contains color material in water. The colors are red, yellow, green and blue. You need only a few drops to color a liter of water. Blend these colors as indicated, and you can produce almost all the colors of the rainbow.

If a demonstration involves water or dilute solutions, you may try adding a few drops of food coloring. The demonstration then becomes decorative, attractive and mysterious.

The Stage

An ordinary table can be used for demonstrations at service clubs or parent-teacher groups if you make the proper preparations. Since there is no bunsen burner, you use an alcohol lamp or propane blast burner for heating purposes. A large beaker of water or a pailful of water should be placed close at hand. The table should be covered with a metal or plastic sheet to protect the table top against flammable or caustic materials.

If the table is small, have another nearby on which you can place materials for the experiments as you perform them. Each chemical, beaker or piece of apparatus should be arranged so that it will be ready for use at the proper time. These materials should be in place before demonstration time so there is no fumbling or hunting for apparatus once the demonstrations have started. Although a public presentation can be done very well by one person it is often helpful to have an assistant who will have materials ready when needed.

As many as four or five persons can put on a chemical magic show with one person acting as chief magician. Dressed in top hat and tails, he can introduce each of the lesser magicians who in turn perform feats of magic. Each performer will master two or three demonstrations and the accompanying patter and stories that go along with them. This type of performance is very effective if it is well planned and well rehearsed.

Visibility

The demonstrations to be visible from all parts of the room should be conducted with as large apparatus as can be conveniently handled. Large beakers, flasks and ring stands are to be preferred to smaller pieces of apparatus. The materials for demonstrations should be easily visible in all parts of the room; and therefore, an elevated table or desk should be used. One-hundred-fifty-watt reflector floodlights can be attached to ring stands and placed in strategic positions. A painted white or black background may help to improve visibility. Forms constructed of plywood can be made easily and with little expense. There should be a base on the form so that a ring stand can be placed on it.

Darken the room if possible for fires, flares or flames of any color or degree of visibility. Some flames are barely visible in a brightly lighted room.

Floodlights focused on the demonstration table should be wired so that you can switch them on or off during the performance. If this is not possible, have an associate turn off lights at your suggestion.

Colored liquids are highly visible whereas water and most water solutions are transparent. A few drops of food coloring added to a liquid make it easier for the audience to see the demonstration. This is especially true for liquids in motion.

Magic Patter

To be entertaining as a performer of chemical magic, you must be a good storyteller. Your stories and fantastic explanations are as important as the experiment itself. You must learn to speak glibly as you carry out the demonstration and be able to divert the attention of your audience from one phase of your work to another. If the experiment fails to work as you had anticipated, your explanation can easily cover up the failure. Quickly you go on to the next demonstration.

After seeing many students and adults perform chemical magic experiments, the author is convinced that the performer must actually know the exact words that accompany each demonstration. The performance appears to have professional competence only if the words spoken have been well rehearsed in the mind of the demonstrator.

The actual words that accompany your demonstration should be repeated slowly but with conviction. If you perform the experiment "Eating a Candle," you may wish to repeat words as follows:

"Ladies and gentlemen, as a chemist I wish to bring to your

attention some of the unusual scientific discoveries of recent years. An explorer friend of mine, who only last week came out of Equatorial Africa, told me about being lost in the jungle. He said that he lived for weeks with nothing but candles to eat. He let me have one of his candles and I have it here on this candlestick." (At this point you light the candle.) "As you see, the candle produced a bright light at night and sustained him in his travels by day." (Now you proceed to blow out the candle and eat it.)

Should you demonstrate the experiment entitled "Hard Water" you may proceed as follows:

"There has been considerable discussion about the hardness of our city water. I have been told that the new gadgets installed by the city have removed the hardness from the city water. Being a chemist, I have been skeptical of these new devices and I shall show by means of a demonstration that there is still hardness left in the city water." (You now pick up the two beakers.) "I have here in my hands two beakers containing city water. The one in my right hand came from a faucet in the north end of the city and in my left hand, from the south end of the city. I wish to show what happens when the two waters meet. As I pour the northern waters into the southern waters and back again you will notice that the waters harden and solidify. I am sure that you will agree with me that the city water is very hard. Not only that, but I shall place a little of this hard water on the table and test it with a flame from a match. As you see, the city water is not only hard but flammable."

The following patter could accompany the demonstration entitled "Synthetic Gold":

"I have a friend, Bill Jones, a confirmed bachelor who is one of the leading atomic scientists of the nation. Bill, however, being a single man, is a suspicious person and in his spare moments has developed an experiment which he tells me has been very useful to him. He has been able, by means of this experiment, to determine which of his women friends are gold diggers. It was only by great persuasion on my part that he consented to reveal his secret to me. The secret ingredients are here in these two beakers. I shall show you now how it has been possible for Bill, or anyone for that matter, to determine if a woman is a gold digger.

"You see that I pour the ingredients from one beaker into the other and I shall hold this remaining beaker over the head of a woman." (At this point you walk down the aisle to the audience and proceed to hold the beaker over the head of one woman followed by two others. You continue your patter.) "You will notice that nothing happens when I hold the beaker over the head of this woman. Apparently the second woman is not a gold digger. Since nothing happens I am beginning to wonder if by chance I have lost the secret. Once more I shall hold the beaker above the head of this woman." (At the moment of placing the beaker over the head of the third person the solution in the flask becomes gold in color. Careful timing is needed so that the color change occurs at the right moment of the patter.)

The magic patter described here can be varied to suit both the demonstrator or the audience. Your showmanship depends largely on how well you learn your good-humored stories or explanations. The actual demonstration is important, of course, but it will fall flat without a good accompanying story.

I. FOAMS

• Black Foam

Action:
Two 200 ml. beakers are standing on the table. You pick them up, pour a clear liquid from one beaker into the other, which is one-third full of a white powder. Stir well with a stirring rod for a few seconds and then place the beaker on the table. When you place a white cardboard behind the beaker, the material begins to darken and gives off fumes. In a few minutes a black solid will rise several inches above the beaker.

You Need:
About 10 ml. of concentrated sulfuric acid in the first beaker; powdered sugar in the second; stirring rod.

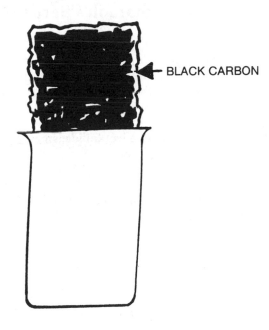

BLACK CARBON

1

Why:

Black carbon remains when the sulfuric acid removes the elements hydrogen and oxygen from sugar. Gases formed cause the material to rise or foam.

Suggestions:

For an instantaneous reaction, try the following experiment. From two 200 ml. beakers, pour two liquids simultaneously into an empty 400 ml. beaker. One contains 50 ml. concentrated sulfuric acid; the other is a concentrated sugar solution made by dissolving 130 grams of sugar in 100 ml. of water. Immediate reaction with considerable frothing occurs when the liquids come in contact. A large plate should be under the beaker to catch the overflow.

Both methods are satisfactory. The first one, however, produces sulfur dioxide fumes that are somewhat suffocating in a small room. The formation of black carbon gives the demonstration a special appeal.

CAUTION: Handle sulfuric acid with great care.

• Pharaoh's Serpent

Action:

Into a small evaporating dish is placed some yellow powder and a few drops of liquid. On slowly heating this mixture a "snake" suddenly leaps out of the dish in a cloud of smoke.

You Need:

Three grams para nitroacetanilide; small evaporating dish; one ml. concentrated sulfuric acid.

Why:

Dehydration is demonstrated. Gas and carbon are formed in the chemical action.

How:

After placing the para nitroacetanilide in the evaporating dish, you add the acid. On heating for two or three minutes, a reaction suddenly occurs and the "snake," which may be over a foot long and several inches in diameter, darts upward.

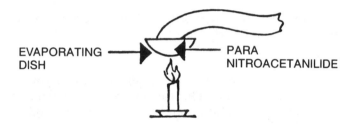

EVAPORATING DISH   PARA NITROACETANILIDE

Suggestions and CAUTION:

The "snake" is composed of carbon. Gases generated in the reaction escape. Some sulfur dioxide gas is formed. Considerable smoke rises to the ceiling at the moment of reaction. This resembles the dome-shaped cloud formed at the explosion of the atomic bomb. The smoke and gas formed in this reaction are irritating to the eyes and lungs. The experiment should therefore be performed shortly before spectators leave the room. In a large room with a high ceiling the fumes and smoke produce little or no irritation.

II. COLOR CHANGES; INVISIBLE INKS

Bloody Picture

Action:

You hold a clear card in one hand and proceed to draw a bloody picture with a finger of your other hand.

You Need:

Five grams potassium thiocyanate; five grams ferric chloride.

How:

Add a few ml. of water to each salt to make saturated solutions. The card is covered with the strong potassium thiocyanate solution. The finger has been dipped in ferric chloride solution.

Why:

Ferric ion reacts with thiocyanate ion to give the red color. This is a sensitive test for the ferric ion.

Suggestions:

You pick up a dagger and thrust it over the back of your hand. You appear to draw blood. The dagger has been dipped in the potassium thiocyanate solution and the back of the hand coated with ferric chloride solution.

Jug of Mystery

Action:

Water is poured from a jug into a series of six empty water glasses. The glasses become filled with liquids colored: (1) red, (2) white, (3) blue, (4) black, (5) green, (6) amber.

You Need:

In the jug, five grams of ferric ammonium sulfate in 500 ml. of water; in each of the glasses, about half a gram of the following solids dissolved in a few ml. of water: (1) potassium thiocyanate, (2) barium chloride, (3) potassium ferrocyanide, (4) tannic acid, (5) tartaric acid, (6) sodium hydrogen sulfite.

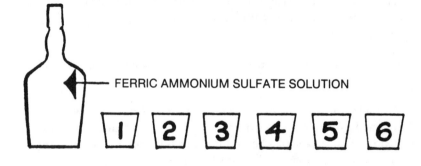

FERRIC AMMONIUM SULFATE SOLUTION

Why:
  (1) Thiocyanate ion forms a deep red color with iron (III).
  (2) Barium ion forms a white cloudy precipitate with sulfate ion.
  (3) Ferrocyanide ion forms a deep blue compound with iron (III).
  (4) Tannic acid forms a black complex with iron (III).
  (5) Tartaric acid forms a greenish complex with iron (III).
  (6) Hydrogen sulfite ion forms an amber product with iron (III).

Suggestions:
  Good lighting helps to make this foolproof experiment effective chemical magic. Use a decorative-appearing jug. Be sure to use the ferric compound, not the ferrous, in the jug. For patriotic colors, use only the first three glasses.

Hot and Cold Colors

Action:
  A pink liquid in a liter beaker stands on the demonstration desk. You heat the liquid and the color fades. On cooling the color returns.

You Need:
  A drop of concentrated ammonia in 500 ml. beaker of water to which has been added a few drops of phenolphthalein.

Why:
  A shift of the equilibrium between ionized ammonium hydroxide to un-ionized ammonia takes place on heating. This change causes loss in color.

Suggestions:
  If you wish to speed up the demonstration use a large test tube which can be heated quickly in a flame and then cooled under the tap.
  If the color does not disappear on heating, you likely have too much ammonia in the solution.

• Invisible Ink

Action:
You place a blank card over a flame. Black letters slowly appear.

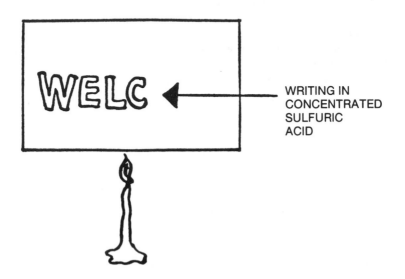

WRITING IN
CONCENTRATED
SULFURIC
ACID

You Need:
A blank card about 6 × 10 inches made of heavy paper; concentrated sulfuric acid.

Why:
Illustrates the dehydrating action of sulfuric acid.

How:
Before the demonstration you will write something on the card. The ink used will be concentrated sulfuric acid and the pen will be a small glass rod.

Suggestions:
Heating the card slowly over the flame will tend to concentrate the acid, remove the elements of water from the paper and leave charred carbon at the points of contact.
Should you wish to use this experiment at the beginning of a

series of demonstrations you may write the word "Welcome" on the card. At the end of a series of demonstrations you may bring your work to a close with the words "That's all" or a similar notation.

CAUTION:
Be careful when working with concentrated sulfuric acid.

Magic Powders

Action:
Two conical piles of white powder of about five grams each are standing on sheets of white paper. To one side is a large cylindrical white box with a cover. You place both powders in the box and shake. Asking the spectators about the color of the powder in the box, you open it. The color is yellow.

You Need:
Five grams powdered lead nitrate; 5 grams powdered potassium iodide; white box.

Why:
Yellow lead iodide is formed by double displacement.

How:
Grind the chemicals separately in a mortar until they are very finely divided. The box must be vibrated very rapidly in the shaking process to get sufficient contact between the chemicals.

Suggestions:
Dropped into a tall cylinder of water the mixed powders give a beautiful yellow colored suspension of lead iodide.

Liquid Thermometer

Action:
A pink liquid in a beaker is standing on the demonstration desk. The color changes to a distinct blue and then back to pink. These changes are repeated continuously.

You Need:
Three grams cobaltous chloride hydrate dissolved in 500 ml. of alcohol.

Why:
The color change is probably due to the shift in the amount of water attached to the ions of cobalt. When warm, the water leaves the ions to be absorbed by the alcohol. Cooling causes a reversal of the process. These changes continue as long as the solution is alternately heated and cooled.

How:
The beaker containing the pink liquid stands on a small hot plate. When the current is on, heat will cause the solution to change from pink to blue. Switching off the current causes a reversal in the color change. A strong light behind the beaker will help to accentuate the color change.

Suggestions:
To make the solution quite sensitive to temperature changes heat it slightly above room temperature. Then add water dropwise until it is pink. The solution will now remain pink at room temperature.

Magical Writing

Action:
A cardboard stands on the demonstration table. It is painted with three colorless solutions. Colors formed will be red, blue and black.

How:
The cardboard has been rubbed with dry ferric chloride. Solutions are potassium thiocyanate, potassium ferrocyanide and tannic acid.

Action:
A painting of a winter scene is shown to the audience. When warmed above a burner, white snow becomes green.

How:
The snow has been painted with cobalt chloride, which becomes bluish-green on warming. You can tell the audience that you are changing the seasons. On painting the blue-green color with water, a pink color returns to the snow.

Action:
Write on a colorless coarse-grained paper with a paint brush dipped in water and the painting is black.

How:
The paper has been rubbed with equal parts of dry tannic acid and ferric ammonium sulfate.

Action:
Write on a colorless coarse-grained paper with a paint brush dipped in water and the painting is red.

How:
The paper has been rubbed with equal parts of dry sodium salicylate and ferric ammonium sulfate.

Action:
Write on a colorless coarse-grained paper with a paint brush dipped in water and the painting is blue.

How:
The paper has been rubbed with equal parts of dry sodium ferrocyanide and ferric ammonium sulfate.

Action:
Using an atomizer, spray a white cardboard with ferric chloride solution. The American flag with all its colors will appear.

How:
An outline of the flag had previously been made with a lead pencil. The stripes had been painted with potassium thiocyanate, the stars with potassium ferrocyanide and the staff with tannic acid solution.

• Mystery Water

Action:
Water from a decorative opaque jug is poured into a series of seven glasses with many peculiar color changes.

You Need:
Jug; seven empty water glasses; five grams tannic acid, a few ml. each of saturated solutions of ferric chloride, oxalic acid, concentrated ammonia and concentrated sulfuric acid.

Why:
Black "ink" results from complex formed of tannic acid and ferric ion (glasses 2 and 4).
In glass 5, the oxalic acid forms a nearly colorless complex with iron (III) by displacement of tannic acid.
In glass 6, ammonia displaces tannic acid–ferric complex to form a yellowish complex.
In glass 7, sulfuric acid destroys this complex to yield a nearly colorless iron (III) ion (hydrated).

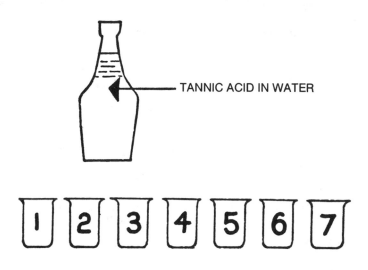

TANNIC ACID IN WATER

How:
Line up the empty glasses in a row on the demonstration table. Into the jug place the tannic acid. Stirring well, fill the jug with distilled water.

Glasses 1 and 3 are left empty.

Glasses 2 and 4 contain five drops of saturated ferric chloride solution.

Glass 5 contains 15 drops of oxalic acid.

Glass 6 contains 10 drops of ammonia.

Glass 7 contains 5 drops of sulfuric acid.

You are now ready for the performance. You pour water from the jug into the first glass—there is no color change—water appears present.

When the liquid is poured from the jug into the second glass, ink appears to pour out.

Poured into the third glass, water appears to pour out.

Poured into the fourth, ink again appears to come out.

You now pour the liquid from all four glasses into the jug.

When poured into glass 1, ink appears to come out.

Poured into number 2, ink also appears to come out.

Poured into number 5, water appears to form.

Poured into number 6, wine appears to form.

When all are poured into the jug, a jugful of wine appears to form.

The wine poured into number 7 appears to form water.

Patriotic Colors

Action:

From a bottle you pour a liquid into each of three beakers standing on a demonstration table. You produce the colors red, white and blue.

You Need:

Solution of alcohol containing phenolphthalein in the first beaker; concentrated lead nitrate in the second beaker; and concentrated copper sulfate in the third beaker. The bottle contains dilute ammonium hydroxide.

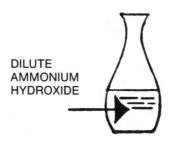

DILUTE
AMMONIUM
HYDROXIDE

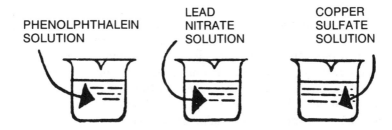

PHENOLPHTHALEIN
SOLUTION

LEAD
NITRATE
SOLUTION

COPPER
SULFATE
SOLUTION

Why:
The action of ammonium hydroxide with the reagents in the beakers produces color changes. In the first beaker, the color change is due to an indicator. Double displacement occurs in the second and a complex ion is formed in the third.

How:
A few drops of reagent in each beaker is sufficient. The intensity of the color depends on the number of drops of reagent used.

Suggestions:
The demonstration has good audience response. It is quite foolproof, and effective with good lighting.

Red and Blue Cloth

Action:
You take a piece of moistened cloth in the hand and dip it into a solution in a beaker. The cloth becomes bright red. Dip into a second beaker and the cloth becomes bright blue.

You Need:
Twenty grams ferric chloride; five grams potassium thiocyanate; ten grams potassium ferrocyanide.

Why:
Two sensitive tests for the ferric ion are demonstrated.

How:
Prepare the three solutions needed for the demonstration by placing each of the chemicals in a separate 400 ml. beaker. Then dissolve the chemicals by adding 100 ml. of water to each. You are now ready to proceed with the demonstration.

Before the performance you moisten the cloth with ferric chloride solution. When you dip the cloth into the potassium thiocyanate solution, the cloth turns red; when you dip it into the potassium ferrocyanide solution it becomes a dark blue.

Suggestions:
The ferric chloride solution poured into the potassium thiocyanate solution turns it a bright red and, when poured into the potassium ferrocyanide solution, dark blue.

Water to Wine to Coffee

Action:
On the demonstration table is a beaker of water. You stir the water vigorously with a glass tube and wine is formed. You place the rod on the table. You now decide to change the wine to coffee. Again you pick up the tube and stir. The wine changes to coffee.

You Need:
Few crystals of potassium permanganate; tannic acid with volume about the size of a small pea; six inches of glass tubing sealed in the middle.

Why:
Water, which becomes wine colored with potassium permanganate, becomes coffee colored in contact with tannic acid.

How:
Previous to the performance you place a crystal or two of potassium permanganate in one end of the tube and tannic acid in the other. Stirring rapidly in the beaker causes the potassium permanganate to dissolve giving the wine color. After placing the tube on the table you stir with the other end causing the tannic acid to react with the permanganate solution giving a coffee color.

Suggestions:
Failure of the experiment may be due to using too large a quantity of the chemicals.

Whiskey to Water

Action:
A whiskey bottle almost full of whiskey stands on the demonstration table. You pick it up, give it a quick shake and the color disappears. The whiskey seems to have changed to water.

You Need:
A large highly decorated whiskey bottle with screw cap; 0.5 gram finely powdered sodium thiosulfate; tincture of iodine.

Why:
Oxidation of sodium thiosulfate by iodine results in a colorless solution.

How:
Add water to the bottle. Into this pour a few drops of tincture of iodine to give it the whiskey color.

The powder is suspended directly below the screw cap of the whiskey bottle. This permits rapid mixing. A satisfactory arrangement can be made from a small sheet of metal and a stiff wire. Shape the metal sheet into a container about 15 mm. long, 5 mm. high and 5 mm. wide. Push the wire through this little metallic cup and then through the cap in such a way that the powder will be suspended about one inch below the base of the cap. Labels over the neck of the bottle will conceal the thiosulfate container.

Powder sodium thiosulfate in a mortar. The powdered salt reacts more quickly than the crystalline form.

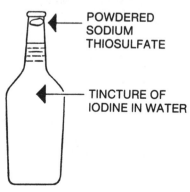

POWDERED
SODIUM
THIOSULFATE

TINCTURE OF
IODINE IN WATER

• Wine to Water to Milk

Action:
You hold up a wine bottle. It is half filled with a liquid that looks like wine. From a Florence flask you pour an invisible material into the wine bottle and wine appears to change to water. The colorless solution is then poured into a milk bottle and this bottle becomes filled with a liquid that appears to be milk.

You Need:
Wine bottle, milk bottle, 500 ml. Florence flask, five grams sodium sulfite, dilute sulfuric acid, few crystals of potassium permanganate, three grams barium chloride.

Why:
Wine-colored potassium permanganate oxidizes sulfur dioxide gas with the formation of sulfate ions in a colorless solution. When poured into the milk bottle the colorless solution forms white insoluble barium sulfate which gives it the appearance of milk.

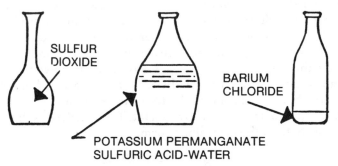

SULFUR DIOXIDE

BARIUM CHLORIDE

POTASSIUM PERMANGANATE
SULFURIC ACID-WATER

How:
Fill the Florence flask with sulfur dioxide. You can generate the gas by the action of a few mls. of dilute acid on the sodium sulfite. Use a large test tube with rubber stopper and delivery tube. Collect the gas by downward displacement. Test with moistened litmus to determine if the flask is filled with gas.

The wine bottle contains 2 ml. of sulfuric acid, a few crystals of potassium permanganate dissolved in water.

The milk bottle contains the barium chloride in a few mls. of distilled water. Add enough water to make a saturated solution.

Suggestions and CAUTION:
Keep the face well away from suffocating fumes of sulfur dioxide. Use a little potassium permanganate to give the wine color. The gas can decolorize only a limited amount.

• Water to Milk to Water

Action:
Three quart milk bottles are standing on the table. The first appears to be half full of water. The others appear to be empty.

You pour the water from the first into the second, changing the water to milk, and the milk formed in the second is poured into the third bottle. Milk formed in the second appears to change to water in the third.

You Need:
Distilled water to make up solutions.
In the first bottle; solution of 1 gram calcium chloride in 500 ml. water.
In the second bottle; solution of 0.2 gram ammonium oxalate in 10 ml. water.
In the third bottle; 5 ml. concentrated sulfuric acid.

Why:
White insoluble calcium oxalate is formed when the first solution is poured into the second. This precipitate dissolves on pouring it into the third bottle.

Suggestions:
This demonstration can also be done by the use of calcium oxide, sodium carbonate and concentrated hydrochloric acid. Place one gram calcium oxide in 500 ml. of water. Stir and filter. This clear solution is placed in the first bottle. In the second bottle place 0.5 gram sodium carbonate in a little water. In the third you place a few mls. of concentrated hydrochloric acid. Pouring the clear limewater which is in the first bottle into the second results in a white precipitate of calcium carbonate. Pouring the contents of the second bottle into the third results in a clear solution since the solid material then dissolves.

Milk can be made to appear to come from water by the use of barium chloride and concentrated sulfuric acid. Dissolve barium chloride in 500 ml. of water in the first bottle. Pour this clear solution into the second bottle containing the acid. An insoluble white precipitate forms which resembles milk.

CAUTION:
Be careful when working with concentrated acids.

• Wonder Picture

Action:
You decide to paint a picture of someone in the audience so you take a sheet of drawing paper and proceed to paint the face of a person. You have two paint pots with a brush in each. The face is painted with one brush and the hair with another. The picture is faint pink and you proceed to warm it over a flame. The face becomes a deep bluish green and the hair a deep violet.

You Need:
A few crystals of hydrated cobaltous chloride dissolved in water in the first paint pot and a few crystals of hydrated cobaltous acetate dissolved in water in the second paint pot.

Suggestions and CAUTION:
A very fine spray of water on the picture changes the color to the original pink.

You can produce an interesting colored picture of a zebra. On a paper draw an outline of a zebra. Draw stripes on the zebra. Use a concentrated solution of antimony chloride to paint every other stripe. The remaining stripes will be painted with a concentrated lead acetate solution. Place the paper with the outline of the zebra in a large jar in full view of the audience. Hydrogen sulfide gas is then passed into the jar. The stripes become alternately orange and black. Hydrogen sulfide gas is poisonous. Do not inhale it.

Acid Breath

Action:
You blow your breath through a straw into a beaker of pink liquid. The liquid turns colorless in a minute or so.

You Need:
Soft drink straw; 250 ml. beaker, half filled with water; 2 to 3 drops phenolphthalein solution; one drop 6 molar sodium hydroxide.

Why:
Carbon dioxide from the breath dissolves in the basic solution, neutralizing it, and turns the indicator colorless.

How:
Add the indicator and the sodium hydroxide to the water and stir. This forms a basic solution which turns the indicator faint pink.

Suggestion:
Do not use too much sodium hydroxide or the carbon dioxide will not be able to neutralize the base and the color will not change.

III. GAS LIBERATION; BUBBLING

• Bubbling Columns

Action:
A pink liquid in a tall cylinder is bubbling vigorously on one end of a demonstration table. A yellow liquid in a similar cylinder located on the other end of the table is likewise bubbling in mysterious fashion.

You Need:
One gram powdered potassium chromate; few crystals potassium permanganate; two tall cylinders or jars; dry ice.

Why:
This is a good demonstration to introduce the magic show. The change of carbon dioxide from solid to the gaseous state is colorfully demonstrated.

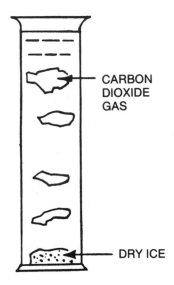

CARBON
DIOXIDE
GAS

DRY ICE

How:
The tall cylinders or large jars are filled with water to within three inches of the top. Stir the potassium permanganate in one cylinder and potassium chromate in the other. You now have pink and yellow solutions.

When the magic show is about to start, you will have in readiness several chunks of dry ice that will be dropped in the solutions. If the chunks are about an inch or two in diameter, they will need replacing in about ten or fifteen minutes. Use a hammer to break the dry ice into the right size.

Suggestions and CAUTION:
Keep the action going throughout the performance of the magic show by adding chunks of dry ice when the action slows down. You must be careful in handling the material. Use paper between the hand and the dry ice or you may freeze your skin.

Educated Moth Balls

Action:
  Little white balls rise and fall in a tall cylinder while spectators are trying to guess the reason for the fascinating motion.

You Need:
  Ten grams marble chips; five grams ordinary salt; dilute hydrochloric acid; moth balls; tall cylinder or beaker; food coloring.

Why:
  Carbon dioxide gas accumulates on each moth ball. In time the gas bubbles will have sufficient buoyancy to lift the moth ball to the surface. Loss of gas at the surface causes the moth balls to sink. This movement continues for hours or days.

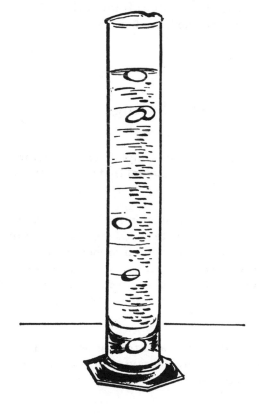

How:

Into the tall cylinder or beaker place the marble chips, salt and 20 ml. of acid. Add a few drops of food coloring and water to within an inch of the top. Drop in several moth balls. After several minutes, they will begin to rise and fall. Stir more salt into the solution if movement does not begin.

Suggestions:

Food coloring, red, green, blue or yellow, merely adds to the beauty of the demonstration.

A cylinder located on each end of the table with movement throughout the performance of a magic show adds to the interest.

You can use this demonstration on a small scale as a centerpiece on the table. Use vinegar and soda to generate the gas.

To obtain a solution of correct density, add salt until the moth balls begin to float. Then add a little more water.

To add interest to the demonstration you may try this variation. Use a knife to shape the moth balls into cubes. With a pencil, put dots on each cube to make them look like dice.

• Fire Extinguisher

Action:

You hold a homemade fire extinguisher in the hand, tip it upside down and shoot a powerful stream that puts out a fire in the sink.

You Need:

Medium-sized side-neck suction flask (17 cm. high) containing ten grams sodium bicarbonate in 300 ml. water; test tube containing concentrated sulfuric acid; solid rubber stopper to fit the flask, clamped tight with a clamp such as used on an old-fashioned soda bottle; thick-walled rubber tube wired to the side neck of the flask; glass nozzle wired to the end of the rubber tube.

How:

The upright test tube of acid resting on the base of the bottle mixes with the bicarbonate solution on inversion. Pressure released forces the liquid out of the bottle.

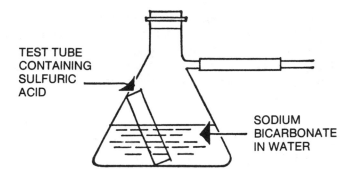

TEST TUBE
CONTAINING
SULFURIC
ACID

SODIUM
BICARBONATE
IN WATER

Why:

Pressure generated by carbon dioxide shoots a stream that blankets a fire.

Suggestions and CAUTION:

Probably the greatest problem in this experiment is to keep the stopper from blowing out. An old-fashioned soda-pop type of clamp is quite suitable for this purpose. You may generate a powerful stream that shoots twenty or more feet out an open window—or you may prefer to direct it into an open jar or pail. Because of the high pressure, you must use care to see that no one is sprayed with the acid solution.

Magical Eggs

Action:

Two large, tall cylinders which appear to contain water stand on the demonstration table. You drop an egg in one cylinder. It sinks but remains suspended half way down. An egg dropped into the other cylinder falls to the bottom but begins to rise in a few minutes, only to fall again. The process of falling and rising continues indefinitely.

You Need:

Two large cylinders; 400 grams salt; dilute HCl solution; two eggs.

Why:

In the first cylinder the egg sinks in ordinary water but remains suspended above the dense salt solution. In the second cylinder, carbon dioxide gas surrounding the egg gives it buoyancy. Loss of gas attached to the egg causes it to lose buoyancy. This occurs when the egg has reached the surface of the solution.

How:

Place the salt in one of the cylinders. With stirring, add water until the cylinder is about one-fourth full. The salt solution should be saturated. Now carefully pour water down the sides of the cylinder until it is nearly full.

To prepare the second cylinder, pour 40 ml. of 6N HCl into it. Add water until nearly full.

Suggestions:

If the egg fails to float above the salt solution, you probably do not have a saturated solution. More salt is needed.

If after some minutes, the egg in the second demonstration does not rise to the surface, add more acid. If it rises to the surface quite rapidly and does not sink to the bottom, you probably have added too much acid. Addition of salt until the egg is nearly ready to float makes it necessary to add very little acid.

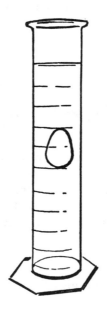

• Soap Bubbles

Action:
Soap bubbles when slipped off a pipe remain suspended several inches above the bottom of a large empty beaker.

You Need:
Large beaker or transparent bowl; Castile soap in a warm mixture of three parts of water to one of glycerine; a carbon dioxide generator or dry ice; pipe.

Why:
Invisible carbon dioxide gas which remains near the bottom of the beaker prevents soap bubbles from sinking to the bottom.

How:
You can generate carbon dioxide by the action of marble chips and acid. You may instead use dry ice to obtain the carbon dioxide. The beaker or bowl should be quite full of the gas.

The soap solution sold in dime stores is good because the bubbles do not break easily. If you make the soap solution with glycerine and Castile soap, the bubbles are quite strong.

Suggestions:
You can produce explosions with soap bubbles. Place a small wad of cotton dampened with gasoline into the pipe and blow bubbles. When you touch the bubble with a lighted match or candle, an explosion occurs. To make a violent explosion blow bubbles with hydrogen gas. If the gas is a mixture of hydrogen and oxygen in the ratio of two to one, the explosion is even more violent.

Soap bubbles slipped off a pipe will remain suspended above the bottom of a large empty beaker if you place a few drops of carbon tetrachloride in the beaker. The heavy vapors of this liquid prevent the soap bubbles from settling to the bottom of the beaker.

CAUTION:
Be very careful when working with gasoline or hydrogen gas.

• Carbon Monoxide

Action:
In a large test tube are placed equal volumes of a solid and a liquid. The gas produced is lighted and burns with a blue flame.

You Need:
Solid sodium formate, 10 grams; 10 ml. concentrated sulfuric acid; match; large test tube, mounted with a clamp on a ring stand.

Why:
Concentrated sulfuric acid reacts with the formate in a dehydrating action which liberates carbon monoxide. The gas burns with a characteristic blue flame.

CAUTION:
Carbon monoxide is poisonous and odorless; the experiment should be performed in a hood. Sulfuric acid is very corrosive and should be handled with extreme care.

• Mysterious Balloons

Action:
Balloons are fastened to the tops of small flasks. When they are raised slightly, they slowly inflate. When released from the flask and closed with a rubber band, the balloons float to the ceiling.

You Need:
Granulated zinc; concentrated hydrochloric acid; rubber bands; balloons; 100 ml. flasks.

Why:
Zinc metal displaces hydrogen from hydrochloric acid. The lighter-than-air gas causes the balloons to rise.

How:

Place some granulated zinc in the balloons. When you raise the balloons, the zinc will fall into the acid, releasing hydrogen slowly. The increase in pressure of the gas causes the balloons to distend. When released from the flask and held upside-down, the balloons are tied off with rubber bands and released.

Remarks and CAUTIONS:

Ordinary rubber balloons will fill satisfactorily if they have been inflated previously. Hydrogen gas is explosive and flames should be kept away. Be careful when working with the acid.

IV. AIR PRESSURE

Blowing Through Glass

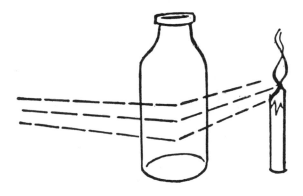

Action:

The demonstrator blows out a candle which is burning behind a wide-mouthed bottle. He appears to blow through the glass.

You Need:

Bottle; candle.

Why:

Bernoulli's principle is illustrated.

How:

Two or three inches behind a bottle is a burning candle. The bottle is approximately three or four inches in diameter. When the demonstrator blows directly at the bottle, he appears to blow through it causing the candle to go out.

Egg in a Bottle

Action:

A rolled up shaft of paper is lit. When burning well you drop the paper into an empty milk bottle. Quickly a wet hard-boiled egg (with shell removed) is placed, with pointed end downward, into the mouth of the bottle. After vibrating upward a few times, the egg is suddenly sucked into the bottle.

To remove the egg, you hold the inverted bottle directly over your head and blow hard. The egg pops out.

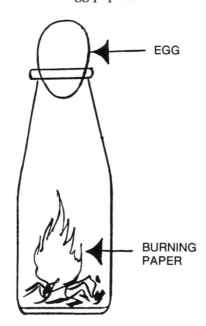

EGG

BURNING
PAPER

You Need:
   Quart milk bottle; hard-boiled egg with shell removed; thin shaft of paper.

Why:
   Heat from burning paper pushes some of the air out of the bottle. A partial vacuum then draws the egg inward. The egg is forced out by blowing hard into the inverted bottle. The inside pressure is increased enough to force the egg out.

Suggestions:
   If the egg with the shell removed has been moistened with water it will slip in and out of the bottle without breaking.
   The thin shaft of paper four inches long should be burning well when dropped into the bottle. The egg should be placed into the mouth of the bottle when heated air is rushing outward.
   To remove the egg, the bottle must be inverted directly overhead.
   To do the experiment more than once, you must fill the bottle with water and empty it. This removes the burned gases and allows fresh air to enter.

Fast Rusting

Action:
   A colored liquid rises in a long glass tube attached to an inverted liter flask filled with steel wool. In ten or fifteen minutes the liquid will ascend into the flask and continue to rise for an hour.

You Need:
   Steel wool; liter flask with one-hole rubber stopper and three feet of glass tubing; crystal of potassium permanganate; dilute hydrochloric acid.

Why:
   Oxygen, combining with iron in steel wool, produces partial vacuum in a flask.

How:
   Over a mass of steel wool about one liter in volume, pour dilute

acid and rinse in tap water. Push this moist steel wool into a one-liter flask. To the flask attach three feet of glass tubing by means of the one-hole rubber stopper. Suspend the arrangement with the flask inverted on a high ring stand over a beaker containing water colored with the potassium permanganate.

Suggestions:

The acid is used to remove rust from the steel wool. The metal with its great surface is oxidized removing oxygen from the air in the flask, resulting in a partial vacuum. This causes the liquid to rise.

The acid treatment should be done shortly before the demonstration since the steel wool oxidizes rapidly after cleaning.

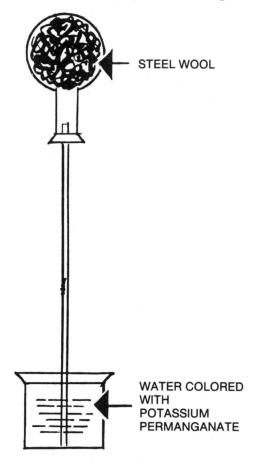

STEEL WOOL

WATER COLORED
WITH
POTASSIUM
PERMANGANATE

Heavy Air

Action:
A one-gallon varnish can crumbles to a shapeless mass.

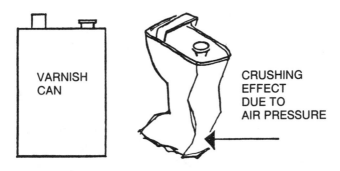

VARNISH
CAN

CRUSHING
EFFECT
DUE TO
AIR PRESSURE

You Need:
An empty one-gallon varnish can with a tight-fitting stopper.

Why:
Condensation of steam in the can causes a partial vacuum. Outside air pressure is not counterbalanced by an equal pressure inside the can which slowly collapses on cooling.

How:
The can is placed over a flame, half a glass of water added and then heated to the boiling point. While still boiling, the can is stoppered with a rubber stopper at the moment of withdrawing the flame. Immediately the can begins to collapse.

Suggestions:
The most common cause of failure is waiting too long to stopper the can after withdrawing the flame. Stopper with a wet stopper while steam is still coming out and the resulting vacuum will make the experiment quite spectacular.

You may wish to try a variation of the demonstration.

Boil 100 ml. of water in a large Erlenmeyer or round-bottomed flask. While still boiling insert a wet, tight-fitting stopper in the mouth of the flask as you withdraw the flame. The partial vacuum will cause the water to boil visibly and further cooling with cold water or ice increases the boiling rate. There is the possibility that the flask may collapse under pressure. Therefore, handle the flask with some care. In a series of magic demonstrations, this experiment is effective since water boils more vigorously when cooled.

Geyser

Action:

A small amount of water is boiling vigorously in a large flask. In the neck of the flask you insert a rubber stopper attached to a long glass tube. Steam now comes from the tube.

You invert the arrangement over a large beaker of water. In a few seconds a powerful spray of water begins to strike the top of the inverted flask with a sound audible throughout the room. Water will almost completely fill the flask.

You Need:

Three-liter round-bottom flask; number 10 one-hole rubber stopper into which is inserted four inches of glass tubing drawn to a point at one end; rubber tubing attached to a four-foot length of glass tubing.

Why:

Air pressure forces water into the flask. Cooling of steam in the flask results in a partial vacuum.

Suggestions:

If the end of the glass tubing that extends through the stopper into the flask has been drawn to a small opening, the force of the fine spray becomes audible. To the other end of the tube you attach the long glass tube by means of rubber tubing.

Before inverting the flask be certain that the small amount of water in the flask is boiling and the steam is emitted from the long tube. Then work fast.

You may wish to have a stepladder so that you can hold the inverted flask high over the beaker.

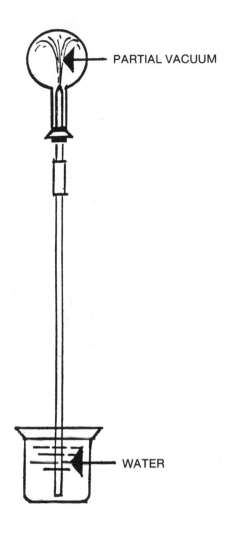

PARTIAL VACUUM

WATER

• Mystery Fountain

Action:

A red-colored liquid rises in a glass tube from a two-liter flask to a similar inverted flask, which is directly above it. The liquid sprays with audible force into the upper flask and changes to a blue color.

You Need:

One meter length of 6 mm. glass tubing; one number 7½ one-hole and one 7½ two-hole rubber stoppers; two 2-liter Florence flasks; short piece of 6 mm. glass tubing and attached one foot of rubber tubing; an ammonia generator made by using a large test tube half filled with equal quantities of mixed ammonium chloride and sodium hydroxide and fitted with a stopper and delivery tube; 50 ml. litmus solution; 10 ml. dilute hydrochloric acid; large ring stand with clamps.

Why:

Liquid rises to the upper flask because of the great solubility of ammonia gas in a small amount of water, creating a vacuum in this flask. Air pressure therefore from below causes the liquid to rise. The color of litmus in acid and base solutions is demonstrated.

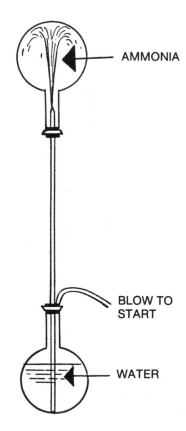

AMMONIA

BLOW TO START

WATER

How:

The six mm. glass tube one meter in length is drawn to a point at one end. This end is lubricated with glycerine and inserted into a number 7½ one-hole stopper to a depth of 10 cm. This in turn is pushed into the upper two-liter flask. The upper flask is known as the ammonia flask.

The other end of the glass tubing extends to the bottom of the lower two-liter flask. It passes through the two-hole stopper. Into the second hole of this stopper attach the short length of glass tubing with rubber tube attached.

Remove the upper flask and pass ammonia gas into it until the dry flask is completely filled with the gas. Ammonia is collected by downward displacement and is generated by heating the test tube. The flask should be completely dry when filled with gas.

Place the litmus solution and acid in the lower flask and fill it with water.

Connect the inverted ammonia flask to the lower flask. Support the arrangement with a ring stand and clamps. Blow into the rubber tube with enough force to cause a few drops of liquid to collect in the upper flask. This starts the fountain.

Suggestions:

Most failures are caused by not generating enough ammonia in the upper flask. When completely filled with ammonia, the liquid will rush upward with enough force to be heard throughout the room.

Hydrogen chloride gas, generated by the action of sodium chloride and concentrated sulfuric acid, is as good as ammonia in the upper flask. Then the lower flask should contain ammonia instead of acid. (*Be careful* when working with concentrated sulfuric acid!)

It is possible to start the fountain without blowing into the rubber tube. You will then use a two-hole rubber stopper in the upper flask. One of the holes will contain a medicine dropper containing water. When you are ready to start the fountain you merely squeeze the medicine dropper, which puts a few drops of water into the flask.

If you do not have litmus solution, use a few drops of other indicators such as methyl orange or phenolphthalein in the lower flask.

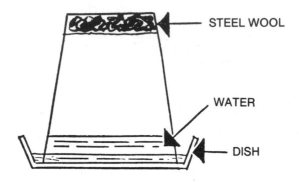

STEEL WOOL

WATER

DISH

Oxygen in Air

Action:
An empty inverted water glass rests on a dish of water. Over a period of several hours water rises in the glass and eventually occupies one-fifth of its volume.

You Need:
Small wad of steel wool; vinegar.

Why:
To show that air is one-fifth oxygen.

How:
Pour vinegar over the steel wool and wedge it into the base of the water glass. Invert over the dish containing water. Rusting of the iron slowly removes the oxygen as the water level rises. A similar, more striking experiment is the one entitled "Fast Rusting."

Obedient Bottle

Action:
A tall cylinder filled with water contains a small inverted bottle floating on the surface. You decide to push the bottle to the bottom of the cylinder. This you do by merely placing the flat of your hand on top of the cylinder. On removal of your hand from the cylinder the bottle rises to the surface.

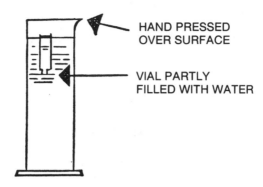

HAND PRESSED
OVER SURFACE

VIAL PARTLY
FILLED WITH WATER

You Need:
Cylinder or tall glass of water; small vial or bottle.

Why:
Air within the bottle is compressed when the hand is placed over the cylinder. This makes the bottle less buoyant and causes it to sink.

How:
Invert the small empty bottle in the cylinder which is then filled with water. Tipping the cylinder over a sink, water can be made to enter the bottle as air escapes until the inverted bottle barely floats. This adjustment must be made carefully since the demonstration fails to work if you have either too much or too little water in the bottle. Now fill the cylinder with water almost to the top.

Pressure of your hand on the top of the cylinder compresses air above the water. This pressure is transmitted through the water to the air enclosed in the bottle. Compression of air in the bottle decreases the buoyancy, causing it to sink. Release of pressure allows air in the bottle to expand giving it enough buoyancy to again rise to the surface.

Syphon Fountain

Action:
Water spraying upward against the base of a large inverted flask rises through a glass tube from an elevated beaker. The water falls to the neck of the flask, then into a glass tube that extends to a lower beaker.

You Need:

Two-liter flask; two large beakers; number 8 two-hole stopper; glass tubing.

Why:

Water is made to rise higher than its source in a syphon arrangement. After spraying with force against the base of an inverted flask the water seeks a lower level.

How:

Glass tubing is drawn to a point producing a small opening. Cut off a section two feet long and insert into the stopper with the

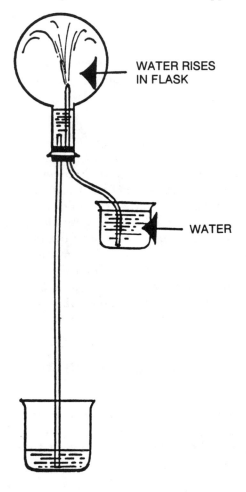

WATER RISES
IN FLASK

WATER

pointed end inside the flask. The other end terminates in the upper beaker. Into the other hole of the stopper insert glass tubing that extends from the flask to the large lower beaker.

To start the syphon, fill the flask with water, insert the rubber stopper and tubing arrangement. Invert and attach to a ring stand on the table. Allow air to enter a little at a time from the tube that dips into the upper beaker. When most of the water has emptied out of the flask the syphon will operate as long as there is water in the upper beaker.

Force of the spray depends on the difference in levels between the beakers. Increase visibility of water movement by adding food coloring to the upper beaker.

Substitute rubber tubing for much of the glass tubing used in this demonstration. You may then move the beakers around more easily.

Upside-Down Water Glass

Action:
An inverted water glass stands on the demonstration table over a sheet of paper.

Another full inverted water glass stands over a similar full one.

You Need:
Three similar water glasses.

Why:
Air pressure prevents the escape of water from the inverted glasses.

How:
Fill the first tumbler full of water, place a sheet of paper over it and carefully invert on the table. The tumbler cannot be removed without upsetting the water.

The second demonstration is similar to the first except the tumbler is brought down carefully over the upright one. Slide the paper out when the open ends of the tumblers coincide.

This parlor trick is one that can be performed at home.

V. BOILING LIQUIDS AND VAPORIZATION

Boiling Water in Paper

Action:
Water is heated to the boiling point in a box-like paper container placed on a screen. The screen supported by a ring stand is above a bunsen burner.

You Need:
Sheet of typewriter paper; four paper clips or Scotch tape; ring stand; ring; screen.

Why:
Conduction of heat through the paper is seen to increase the temperature of water to the boiling point.

How:
Fold typewriter (or stronger) paper about two inches inward from four directions and fasten the ends together with paper clips or Scotch tape. The base of this box-like container will be about 6 × 4 inches. Pour in about 200 ml. of water.

Suggestions:
An interesting variation of the experiment is to boil water in a paper bag. Water in contact with the paper absorbs the heat, keeping the temperature low enough to prevent combustion of the paper. Water is heated slowly in these experiments since paper is a very poor conductor of heat.

• Phosphorus Glow

Action:
A phosphorescent glow appears above a glass tube protruding above a large test tube. Boiling water in the test tube is being

heated above a burner. A match head placed in the glow does not ignite. Visibility of the glow becomes marked when the room is darkened.

You Need:

White phosphorus, size of a pea; a large test tube with one-hole rubber stopper into which extends three inches of glass tubing.

Why:

Slow oxidation gives rise to phosphorus glow when the element is vaporized.

How:

Place water to a depth of one or two inches and the phosphorus in the test tube. Attach the stopper assembly and heat to boiling. The phosphorus glow which becomes highly visible in a darkened room continues as long as the water is boiling.

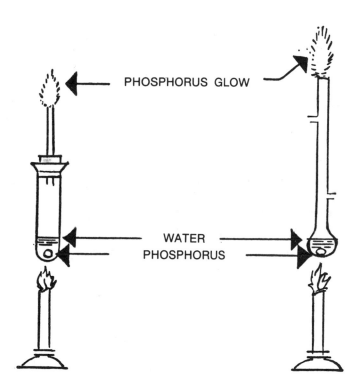

PHOSPHORUS GLOW

WATER
PHOSPHORUS

Suggestions:

The peculiar nature of the ghastly flame that does not burn makes this an excellent magic type of demonstration.

If you wish to make a phosphorus glow four or five inches long set up a reflux condenser without the hose connections. Placing a small piece of white phosphorus in the flask in water, heat to boiling. The phosphorescent glow can be made to exist at various positions along the condenser tube or at its top by varying the size of flame of the burner. To see the glow satisfactorily you need a darkened room.

CAUTION:

White phosphorus causes serious burns.

Cold Boiling

Action:

A flask of water is boiling on a ring stand mount. The flask is removed, quickly stoppered, and placed under a cold water tap. The water in the flask continues to boil furiously for several minutes.

You Need:

One liter spherical flask (Pyrex); ring stand and clamp; rubber stopper.

Why:

When boiling the flask is full of steam which rapidly condenses under cold water. At reduced pressure the water will boil at lower temperature.

Drinking Bird

Action:

A glass toy bird is perched on the lip of a drinking glass. You dunk the bird's head in the water in the glass; the bird rises to erect position, then continues to take a sip occasionally.

You Need:
"The Amazing Drinking Bird," model available from Edmund Scientific Co., Barrington, New Jersey 08007, telephone (609) 547-3488 (item #60778); water glass or cocktail glass; water.

Why:
When the bird's head is dry, the center of gravity is below the pitch axis, and the bird sits erect. When the bird's beak is wet and the water begins to evaporate, cooling occurs. The volatile liquid sealed inside the bird has a vapor in equilibrium. This vapor condenses, causing a partial vacuum. Liquid then is drawn up the body of the bird into the head, and the bird pitches forward, again wetting the beak. Gravity draws the liquid from the head into the body, and the bird rises erect again. The cycle continues indefinitely.

How:
First, wet the head completely. Put the pivot on the wire first, then hang the wire over the lip of the glass.

Suggestions:
Keep the water in the glass constantly leveled and cool. Do not drip water on the bird's tail. The demonstration works best where air flow is good.

VI. FIRES AND COMBUSTION

• Blue Flare

Action:
A few drops of water from a medicine dropper fall on a small mound of powder. An instantaneous blue flare is accompanied by smoke.

You Need:
Four grams powdered ammonium nitrate; one gram powdered ammonium chloride; zinc dust.

Why:

The catalytic effect of water is shown on a mixture of oxidizing and reducing agents.

How:

SEPARATELY grind the chemicals in a mortar. Mix the ammonium nitrate and ammonium chloride and place in a mound on a metal sheet. Sprinkle with zinc dust. You are now ready for the reaction with water.

CAUTION:

Fumes and smoke that accompany this reaction limit its use to a large auditorium or a well-ventilated room. The high temperature accompanying the reaction suggests CAUTION.

Burning Sugar Lump

Action:

You challenge members of the audience to light a sugar lump with a match. You pass out sugar lumps and matches. No one is able to make a sugar lump burn. It merely melts when fire from the match comes in contact with it. You ask one of the spectators to pass a sugar lump back to you. When you set fire to the sugar lump with fire from a match, it burns with a flame.

You Need:

Sugar lumps; matches; cigarette ashes.

Why:

Cigarette ashes act as a catalyst in causing the sugar to burn.

How:

On receiving the sugar lump from a spectator you push it against cigarette ashes which you have in your hand, or lying on the table. You light the sugar lump at the point of contact between the ashes and the sugar. It catches fire and burns at this point.

Candle Tricks

I

Action:

A candle burns on the demonstration desk. Over its center you place a wire screen. The flame now burns below the screen. By sliding the wire to one side and back again to the center, the flame can be made to burn partly above the screen and partly below.

By gently raising and lowering the screen slightly, you can cause the flame to burn entirely above the screen. The small flame above the screen can be made to rise and fall as much as three inches.

You Need:

Candle; screen.

Why:

The loss of heat to the screen is shown.

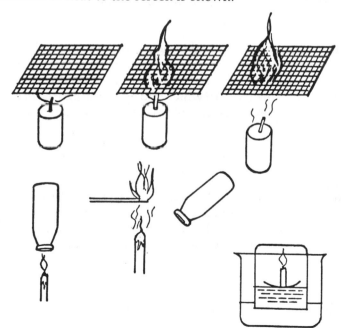

Suggestions:
   If you are willing to spend half an hour in practice with the candle and screen, you will learn how to adjust the distance between the candle and screen so that the burning gas will rise a considerable distance above the candle. Moving air currents tend to disrupt the demonstration.

II

Action:
   Holding an inverted empty half pint cream bottle in one hand you place it over a burning candle. The candle is extinguished. Quickly you lift the bottle upward and bring a lighted match to the extinguished candle. The stream of rising gas ignites as the candle is lit.

You Need:
   Candle; match; half pint cream bottle.

Why:
   Hydrocarbon vapors are shown to be combustible.

Suggestions:
   Ignition of gas at a considerable distance above the candle makes the experiment somewhat striking. You must not allow moving air currents to disturb the upward motion of the heated vapor above the candle. The rapidity with which you bring the lighted match to the hot vapors determines the success of your demonstration.

III

Action:
   A small candle burns in the center of a watch glass, which is floating in a large beaker half-filled with water. An empty smaller beaker is inverted over the larger one. Water rises in the smaller beaker as the candle is extinguished. Limewater added to the water turns milky.

You Need:
   Two beakers, 600 and 2000 ml.; 3-inch watch glass; small candle; limewater.

Why:

Part of the air is used up in combustion. This is evidenced by rising water and the fact that the flame is extinguished. The limewater reaction shows that carbon dioxide is formed in combustion. Condensation of moisture on the base and walls above the flame is evidence of oxidation of a hydrocarbon.

Suggestions:

To make limewater, add a small amount of calcium oxide to distilled water and filter.

A slight variation of the demonstration can be made. Attach a burning tall candle in the center of a large empty upright beaker. Add water to half the height of the candle. On inverting a bottle or cylinder over the candle, water rises as the candle is extinguished.

• Burning Water

Action:

From a water pitcher on the table you pour yourself a glass of water, drink some and empty the rest into a pneumatic trough. The water appears to catch fire and burns over the entire surface.

You Need:

Potassium metal (size of a pea) wrapped tightly in filter paper; 10 ml. ether; pneumatic trough; water pitcher; glass.

Why:

The startling demonstration shows that potassium is very active and ether highly flammable.

How:

The potassium metal tightly wrapped in filter paper is placed in the empty pneumatic trough. Being careful that no flames are near, pour the ether into the trough. You are now ready for the demonstration. Water in the trough causes the ether and filter paper to float above its surface. Potassium metal reacts violently with water, generating hydrogen and releasing heat which ignites the ether. Flames may rise two feet in the air.

CAUTION:
Ether is EXPLOSIVE near a flame. Be certain that no flames are near when pouring ether from one container to another. The violence of reaction between potassium and water may produce spattering. Stand to one side.

Cold Fire

Action:
Soak a handkerchief in several ml. of liquid. Ignite the handkerchief with a match. The handkerchief burns with a blue flame but is not damaged.

You Need:
A mixture of 50 ml. of isopropyl alcohol and 50 ml. of water.

Why:
Cooling by rapid evaporation prevents burning of the cloth. What is burning is alcohol vapor.

Suggestion:
The demonstration is especially effective in a darkened room.

• Glowing Steel Ball

Action:
You roll steel wool into a ball about half the size of a golf ball. A stoppered bottle stands to one side of the table. With tongs, the ball is placed in a flame. After the ball begins to glow you blow your breath on the burning ball with a blowpipe or glass tube. With the tongs, you lower the glowing ball into the empty bottle. Brilliant sparks are produced.

You Need:
Steel wool; tongs; stoppered wide-mouthed bottle; blowpipe; metal sheet; an arrangement for generating oxygen, or an oxygen cylinder.

Why:

Intense fire on rapid oxidation of finely divided steel wool produces magnetic iron oxide.

How:

Previous to the performance you fill the bottle with oxygen. Then stopper it.

Suggestions and CAUTION:

Combustion of steel wool in oxygen produces a high temperature. There is some danger of the bottle breaking. Use the metal pad under the bottle. Be careful when working with oxygen.

• Disappearing Flame

Action:

On the demonstration desk are a burning candle and two large stoppered graduates.

Removing the stopper from the first graduate you pour an invisible gas over the flame which is then extinguished. Pouring an invisible gas from the second graduate you relight the flame of the candle.

You Need:
Source of carbon dioxide gas and oxygen gas; two 500 ml. graduates; two number 10 rubber stoppers; candle.

Why:
Oxygen is seen to promote combustion whereas carbon dioxide has the opposite effect and will cause the flame to go out.

How:
Carbon dioxide gas is in the first graduate. This gas can be generated by the action of dilute hydrochloric acid on marble chips. The second graduate contains oxygen gas which was generated by the action of heat on potassium chlorate mixed with a little manganese dioxide.

Suggestions and CAUTION:
After the candle has been extinguished by the carbon dioxide gas, you must pour the oxygen gas over the wick immediately.

The cylinders should be completely filled with gas and tightly stoppered. If the experiment fails, it is likely that you do not have enough gas in the cylinders. If you use bottles in place of cylinders, they should have large mouths so that the concentration of gas will be high at the moment of pouring over the candle.

Be careful when working with oxygen.

• Flare

Action:
One drop of water from a medicine dropper falls on a small cone of powder. An instantaneous bright flare and smoke are followed by a glowing mass that persists for a couple of minutes.

You Need:
Five grams powdered aluminum; 0.5 gram sodium peroxide.

Why:
High temperatures accompany oxidation of aluminum by peroxide.

How:

On a metal mat place a cone of the powdered aluminum to a height of 0.5 inch. Sprinkle the sodium peroxide loosely over the metal and mix it slightly into the metal.

Suggestions and CAUTIONS:

Addition of a drop of water to the sodium peroxide generates oxygen. The heat of reaction is great enough to cause the powdered aluminum to burn with an intense flame which is blinding to the eye. After the initial flare the metal continues to glow. Sodium peroxide is somewhat difficult to handle and materials after the combustion should be flushed down the sink. Great care must be taken to guard against burns since the reaction is rapid, the heat intense and the products corrosive.

Eating a Candle

Action:

A lighted candle on a candlestick is burning. You pick up the candlestick, blow out the candle and quickly proceed to eat it.

You Need:

Candlestick with drippings; banana; pecan nut.

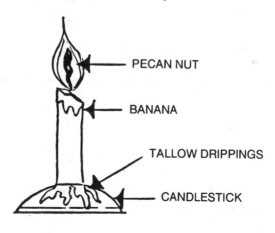

PECAN NUT

BANANA

TALLOW DRIPPINGS

CANDLESTICK

Why:
   The flame from burning oil in a nut resembles that of a candle.

How:
   The candlestick should preferably have some tallow drippings on it to make it look realistic. The banana should be shaped into the form of a candle and be of a size that it can be eaten in a mouthful or two. The pecan nut will be shaped to resemble a wick and inserted in the top of a banana. Since the nut has high oil content, it will burn like a candle for several minutes.

Suggestions:
   A rather small piece of banana is used so that all of it will be eaten.

• Ether Fires

Action:
   An eaves trough rests diagonally between the edge of a table and the floor. You pour ether over a small towel that is resting near the top of the trough. When you bring a lighted candle to the bottom of the trough a large flame rushes upward igniting the towel. You wrap a larger towel around the flaming towel to extinguish the flame.

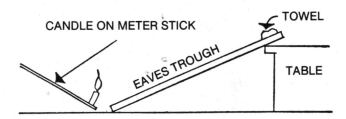

You Need:
   25 ml. ether; 10-foot section of eaves trough; candle; towels; meter stick.

Why:
   The demonstration emphasizes the flammability of a highly volatile liquid and the high density of ether vapor.

How:

The candle used for ignition should be burning on the end of a meter stick. This avoids the danger of ignition near the hand.

Suggestions and CAUTION:

DANGER! The demonstration should be performed by someone fully aware of the highly explosive nature of ether. Liquid ether should never be near an open flame.

Have a fire extinguisher near.

• Fire Spray

Action:

You light a fuse protruding from a hole in an inverted iron crucible. When the fuse burns down, an eruption occurs that shoots a fiery spray many feet in the air.

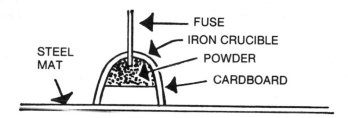

You Need:

Equal parts of powdered magnesium, powdered zinc, powdered iron, powdered charcoal and powdered sulfur, and a double portion of powdered potassium nitrate.

Why:

A violent reaction occurs on ignition of mixed oxidizing and reducing agents.

How:

Make a hole in the base of an iron crucible. Add the powder which has been well mixed in a tin mixing can. The powder can be retained in the base of the crucible by inserting cardboard. Invert the crucible and place the paper fuse in the hole.

The fuse can be made from filter paper which has been soaked in saturated potassium nitrate and dried.

To ignite the mixture, place the crucible on a steel mat, inverted preferably out of doors and away from spectators.

Suggestions and CAUTION:
Each of the materials should be dried separately, ground separately in a mortar and then thoroughly mixed. Caution is advised.

Fireworks

Action:
On the demonstration table are small piles of powder into which has been placed a thin taper of paper. Ignition of the paper will in turn cause the powder to flare up with the colors ordinarily seen in fireworks.

You Need:
A mixture of chemicals, each ingredient in powder form, mixed in the ratios indicated.

BLUE FIRE
potassium chlorate 8
copper sulfide 2
sulfur 4
mercurous chloride 2
copper oxide 1
charcoal 1

GREEN FIRE
barium nitrate 12
potassium chlorate 3
sulfur 2

WHITE FIRE
potassium nitrate 7
antimony sulfide 1
sulfur 1

RED FIRE
strontium nitrate 4
potassium chlorate 4
charcoal 2
sulfur 1

YELLOW FIRE
potassiuim chlorate 6
sodium oxalate 2
charcoal 2
sulfur 1

PURPLE FIRE
copper sulfate 1
sulfur 1
potassium chlorate 1

How:

Each substance should be ground to a powder separately in a mortar, dried, then placed on a large sheet of paper in the ratios indicated. The mixing is done by rocking the paper back and forth. The small pile of mixed powder is placed on a metal mat for ignition. Ignite by placing a thin piece of filter paper in the pile and lighting it with a match. To make the filter paper sensitive soak it first in a concentrated solution of potassium nitrate. Then allow it to dry.

• Fire Water

Action:

You pick up a glass, appear to drink some of the liquid and spit it out. The liquid falls on the table with a great burst of flame.

You Need:

Glass containing ethyl alcohol, few grams dry, red, chromic anhydride.

Why:

The powerful oxidizing agent chromic anhydride reacts with alcohol. Heat generated results in rapid combustion of the alcohol.

How:

On a metal sheet on the demonstration table scatter a few crystals of chromic anhydride. When the alcohol strikes the chemical, an immediate reaction gives rise to flames that rise a foot or more in the air.

Suggestions and CAUTION:
    Try this variation of the experiment. Place some crystals of the chemical in a 500 ml. flask. Now add a few ml. of alcohol. The resulting reaction causes a fire to burn in the flask with a greenish glow. DANGER! All fire experiments require caution to protect the demonstrator as well as the audience. Alcohol must be expelled from the mouth very soon after sipping it or it will cause dehydration.

Fire Writing

Action:
    You touch the lighted end of a cigarette to one side of a sheet of paper. The word "Welcome" is gradually spelled out in fire across the paper. The paper is consumed only at the point of burning.

CIGARETTE

TRACED WITH POTASSIUM NITRATE SOLUTION

You Need:
    Ten grams potassium nitrate in 25 ml. water; small paint brush; fairly heavy paper that is somewhat absorbent.

Why:
    Burning paper is oxidized by potassium nitrate.

How:
    Paint the word on the paper with the saturated potassium nitrate solution. You should go over the word more than once to get enough of the salt in contact with it. Be certain that all the letters

are connected or the fire will go out. The paper must be dry when lighted. Mimeograph paper seems to work well.

Suggestions:
You may wish to burn out other words than the one suggested or you may wish to draw out pictures of animals or other objects. The experiment is easy to perform and shows off best in the dark.

Glowing Splint

Action:
You heat a test tube one third full of white powder. On dropping an unlighted wooden splint into the molten liquid there is a sudden burst of flame and smoke. Use a forceps to drop the splint.

You Need:
Five grams powdered potassium chlorate; test tube; wooden splint.

Why:
Wood burns rapidly in a hot atmosphere with oxygen.

How:
Heat the powder to a temperature high enough to melt the substance. The oxygen generated at this temperature quickly ignites the wood. Since there may be some spattering, be careful.

Suggestion:
Use a "life-saver" or lump of sugar instead of a wooden splint.

• Fire in the Water

Action:
Brilliant flashes of fire burst out at intervals under the surface of water in a beaker. The fire occurs at the point of contact of a bubbling gas.

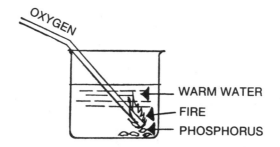

You Need:
One gram white phosphorus; source of oxygen gas; 400 ml. beaker.

Why:
Warm phosphorus in contact with oxygen burns rapidly.

How:
Heat 200 ml. of water in a 400 ml. beaker to 70 degrees C. Place in the beaker several small pieces of white phosphorus. CAUTION: The phosphorus will melt in the water.

Oxygen, which can be generated by heating a mixture of potassium chlorate with manganese dioxide, is bubbled rapidly into the beaker. Flashes of fire occur at the point of contact between phosphorus and oxygen.

Suggestions and CAUTION:
Water is warmed to increase reaction rate. There is some danger that the pieces of phosphorus may explode on ignition but the small amount used makes the experiment quite safe. Smoke given off is not particularly disagreeable. Phosphorus burns are serious. Use great care in this experiment. Handle phosphorus with a forceps.

• Fire Wand

Action:
You bring the end of a glass rod in contact with the wick of a candle. There is a flare and the candle is lit.

You Need:
Equal quantities of powdered potassium chlorate and sugar; a large candle whose wick has fluffy fibers.

How:
The end of the glass rod has been touched with concentrated sulfuric acid. This in contact with the mixed powder on the wick produces an instantaneous flare which lights the wick.

The wick should be ruffled and the powder well interspersed in it. This will insure continued burning after the wick has been touched.

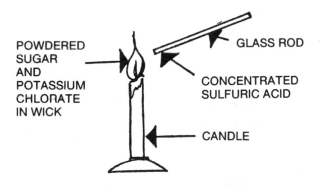

POWDERED SUGAR AND POTASSIUM CHLORATE IN WICK

GLASS ROD

CONCENTRATED SULFURIC ACID

CANDLE

Why:
Combustion of sugar is rapid in the presence of potassium chlorate.

CAUTION:
Grind the crystals of potassium chlorate and sugar separately in a mortar. If mixed and then ground, you may have an explosion.

Glowworm

Action:

A coiled wire suspended from a cardboard cover over a 400 ml. beaker continues to glow as you carry the arrangement about the room.

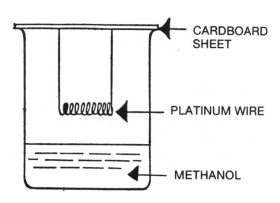

CARDBOARD
SHEET

PLATINUM WIRE

METHANOL

You Need:

Ten cm. of number 27 platinum wire; 100 ml. methanol; 400 ml. beaker.

Why:

Oxidation of methanol to formaldehyde takes place in the presence of platinum, which serves as a catalyst. Heat from the reaction keeps the wire red-hot.

Construction:

A thin circular cardboard sheet with two openings serves as a cover for the beaker. From the center of the cardboard is suspended a 10 cm. length of number 27 platinum wire. The lower end of the wire is coiled so that it is suspended just above the wood alcohol which has been poured into the beaker to a depth of about 5 cm.

How:

Heat the wire to redness in a bunsen flame. Quickly lower the wire and cover to the beaker. The wire will then continue to glow for hours.

Suggestions:

Since the platinum cools off rapidly the wire must be lowered to the beaker as quickly as possible. After glowing for a time above the alcohol the liquid becomes warmed hot enough to ignite. The flame, however, will quickly die out but the ignition will be repeated many times.

If the demonstration does not work, try warming the alcohol and lower the wire rapidly after making it glow in the flame.

The wire will glow for hours and the demonstration is an excellent one to show in a darkened room. The smell of formaldehyde is detected after a time.

Hand warmers used by fishermen and hunters operate on the principles shown in the demonstration.

Concentrated ammonium hydroxide can be used in place of methanol in this demonstration.

Obedient Candle

Action:

You hold a burning candle in one hand and a lighted match in the other. Blow out the candle with a quick puff and quickly bring the lighted match slightly above the wick but not close enough to touch it. The candle lights with a quick flare.

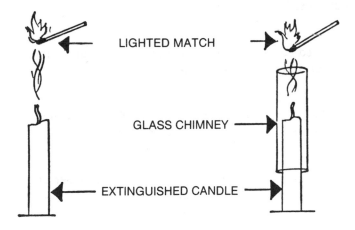

LIGHTED MATCH

GLASS CHIMNEY

EXTINGUISHED CANDLE

Why:
The flame of the match is in contact with heated hydrocarbon vapors. This causes immediate combustion.

Suggestions:
Rapid lighting after the candle is extinguished is needed to make the demonstration successful.

Keep drafts away from the candle.

To make the experiment quite foolproof, try the following variation using a chimney. Fasten the candle to the base of a ring stand. Use a 6- or 8-inch length of glass tubing as chimney. It should be a little greater in diameter than the candle. Clamped to the ring stand, the chimney extends two or three inches above the wick but permits air to enter from below. When the candle is blown out, vapors rise and can be lit above the chimney.

Delayed Fire

Action:
Several drops of liquid from a medicine dropper fall into a paper cup placed over a metal mat. After several seconds a reaction takes place with a burst of flames.

You Need:
One gram pulverized potassium permanganate; glycerine in a small dropper bottle; paper cup; metal mat.

Why:
Pulverized potassium permanganate oxidizes glycerine rapidly. Heat generated results in a flame.

How:
Place the paper cup containing the potassium permanganate on a mat. Glycerine is then dropped into the cup.

Suggestions:
You may use an iron crucible in place of a paper cup. If the crucible has been warmed previously, an immediate reaction takes

place. Otherwise, a period of up to a minute will elapse before the flame occurs. This experiment is most effective in a darkened room.

• Self-Lighting Candles

Action:
Candles placed on a narrow board on the demonstration table will smoke and ignite at various intervals of time.

You Need:
A solution of two grams of yellow phosphorus dissolved in five times its volume of carbon disulfide; a dozen Christmas candles.

Why:
Evaporation of the solvent leaves phosphorus in a finely divided state on the candle wick. The phosphorus then ignites at room temperature.

How:
Arrange the candles upright in a row on a narrow board. Place this on the demonstration table. Ordinary Christmas or birthday candles which have not previously been lit are satisfactory. The phosphorus solution, kept in a dropper bottle, is used to moisten the wicks of the candles ten or fifteen minutes before they will be expected to ignite. Use only a fraction of a drop on each candle.

CAUTIONS:
Keep the dropper bottle containing the phosphorus solution inside of a large wide-mouthed stoppered bottle as a precaution against spilling the highly flammable liquid. A drop of solution spilled on anything combustible will ignite.

Yellow phosphorus is dangerous to handle since it may ignite when you are attempting to cut it. You should cut phosphorus under water and handle it with forceps and not with the hands. Phosphorus burns are serious. This demonstration must be conducted by a person who has the experience and scientific knowledge needed to work with phosphorus.

• Rat Nest

Action:
The demonstrator drinks water from a glass. He decides to sprinkle some of it on a "rat nest" with a medicine dropper. Immediately the rat nest bursts into a vigorous flame and burns despite the water on it.

You Need:
One gram sodium peroxide; excelsior (tinder or wood shavings); evaporating dish; medicine dropper.

Why:
Oxygen is formed when water reacts with sodium peroxide. Heat that accompanies the reaction results in rapid combustion.

How:
On a small ball of excelsior in the evaporating dish, sprinkle the sodium peroxide. Drops of water from the medicine dropper on the sodium peroxide start a rapid reaction.

CAUTIONS:
Sodium peroxide is caustic in contact with water. The intense fire produced in this demonstration could break the evaporating dish. Smoke is produced. Sodium peroxide should be purchased in small amounts and kept air tight in its original container.

Spontaneous Fires

• I

Action:
A few drops of liquid from a medicine dropper fall on a mound of powder, in the center of which a small depression has been made. A violent fire flares up.

You Need:
Granulated sugar; powdered potassium chlorate; concentrated sulfuric acid.

Why:
Dehydration and oxidation of sugar is accompanied by flames.

How:
Powder the chemicals separately in a mortar. Place equal volumes of the mixed materials in a mound on an insulated mat. When a few drops of acid fall on the mixture, a reaction produces an immediate fire.

CAUTION:
Be careful when working with fire and strong acid!

• II

Action:
One drop of water from a medicine dropper falls on a small cone of powder. An intense fire burns. This is followed by a glowing mass for a minute or two.

You Need:
Powdered aluminum; sodium peroxide.

Why:
Addition of a drop of water to the sodium peroxide generates oxygen. Reaction of the gas with powdered aluminum produces aluminum oxide. The heat generated is great enough to cause the powdered aluminum to burn with such an intense flame that the flash is blinding. After the initial flare the metal continues to glow for some time.

How:
On a metal mat place a cone of aluminum to a height of one-half inch. On top of the metal place a small volume of sodium peroxide. A volume the size of a pea is sufficient.

Suggestions:
Sodium peroxide is somewhat difficult to handle. Materials after combustion should be flushed down the sink.

CAUTION:
Great care should be taken to guard against burns. The reaction is rapid and the heat is intense.

• III

Action:
A small evaporating dish is being heated by a flame. A black powder dropped in the heated dish produces brilliant sparks.

You Need:
Two grams potassium nitrate; powdered charcoal.

Why:
Heated potassium nitrate decomposes to yield oxygen. Powdered charcoal burns instantly in the presence of this gas at the high temperature of the experiment.

How:
When potassium nitrate has been heated until it is molten, carefully drop in the powdered charcoal. The reaction is brilliant and instantaneous.

Suggestions and CAUTION:
Be careful, since some burned carbon may be scattered.
Grind ordinary charcoal in a mortar to prepare powdered material.

• IV

Action:
Standing on a chair with a sealed test tube in your hand, you remove the cork. As you sprinkle the contents of the tube in the air, they catch fire in a spectacular display.

You Need:
Five grams ferrous oxalate; paraffin; test tube with cork to fit.

Why:
Finely divided particles of iron and carbon ignite on exposure to air.

How:
Previous to the performance you prepare one or more test tubes to be used in this demonstration. Heat ferrous oxalate in the test

tube until no more fumes are given off. As the test tube is being heated, you melt paraffin in an evaporating dish. Place the cork in the melted paraffin. While the test tube is still hot, pick up the cork with tongs. Seal the tube. On cooling, the melted paraffin will make an airtight seal.

Suggestions:
This fire display is especially attractive in a darkened room.
Five grams lead tartrate can be used in place of ferrous oxalate. Heat the white powder in a test tube until it is black. Then seal the tube with a cork which has been dipped in melted paraffin.

Test Tube Fire

Action:
A large vertical test tube one-fourth full of a white solid is strongly heated until the material melts. Darkening the room, you extinguish the burner and carefully drop in several pieces of charcoal. The room is lighted up with a bright violet-reddish glow. Carbon particles dance about on the surface of the liquid with a popping sound.

You Need:
25 grams potassium nitrate; charcoal; 200 ml. test tube; spoon.

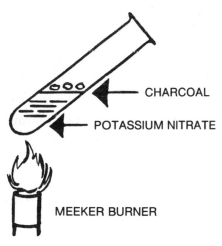

CHARCOAL

POTASSIUM NITRATE

MEEKER BURNER

Why:

Oxygen produced by heating potassium nitrate combines rapidly with carbon. The bright violet-reddish glow is characteristic of potassium.

How:

To heat the test tube rapidly use a Meeker burner. Oxygen liberated at the high temperature of the molten potassium nitrate unites with carbon with such rapidity that a slight explosion seems to occur. To continue the demonstration heat the test tube as you drop charcoal into the molten material.

Suggestions:

Instead of heating potassium nitrate in a test tube you may heat it in a casserole or evaporating dish. Sprinkling powdered charcoal on the molten salt produces a beautiful colored effect.

• Turpentine Fire

Action:

From a medicine dropper you allow a few drops of liquid to fall into a beaker. Flames shoot upward with smoke.

You Need:

A large beaker containing 30 ml. of concentrated sulfuric acid and 20 ml. of concentrated nitric acid; a few ml. of turpentine.

Why:

Rapid oxidation and combustion of turpentine takes place when the liquid is in contact with the acids.

How:

Cautiously mix the acids in the beaker. Hold the dropper containing the turpentine about two feet above the beaker. When drops strike the acid mixture, flames rise to a height of four to six inches.

CAUTION:
Handle the concentrated acids with great care. Fumes and smoke produced make the demonstration one that should be performed with sufficient ventilation.

• Thermite Reaction

Action:
You light a magnesium fuse that starts a spectacular fire. Examination of the container shows that molten iron was formed.

You Need:
Powdered ferric oxide; powdered aluminum; barium peroxide; magnesium ribbon; iron or clay crucible.

Why:
Aluminum unites with the oxygen in ferric oxide. The high temperature of reaction produces molten iron.

How:
On top of a pail of sand place a small iron or clay crucible. Into the crucible place a mixture of three parts of finely powdered ferric oxide and one part of powdered aluminum. Make a slight depression in the center of the surface of the mixed chemicals and in this depression place a small amount of a mixture of ten parts of barium peroxide and one part of aluminum powder. This mixture should make a small mound about a centimeter in height. The fuse which is inserted in this last mixture is magnesium ribbon about ten centimeters in length which has been bent and twisted several times to form a fuse four or five centimeters in length.

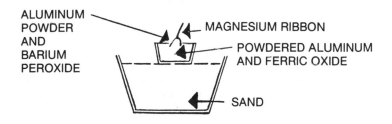

ALUMINUM POWDER AND BARIUM PEROXIDE

MAGNESIUM RIBBON

POWDERED ALUMINUM AND FERRIC OXIDE

SAND

CAUTION:
On igniting the fuse move quickly away since the heat generated is so intense that the whole mass is white hot at the moment of reaction.

Suggestions:
The thermite mixture can be purchased at a supply house.

To make the demonstration more spectacular make a hole in the bottom of the crucible and attach it to a ring stand. The crucible should be about a foot above a jar of water. Sand covers the bottom of the jar. There is a brilliant display as molten iron flows through the hole in the bottom of the crucible, strikes the water and glows above the sand.

Volcano

Action:
Red-hot particles and smoke are seen to erupt in a miniature volcano when you light a conical pile of red powder in a liter beaker.

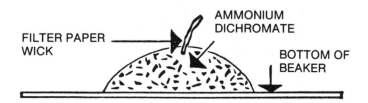

You Need:
20 grams powdered ammonium dichromate; liter beaker; filter paper; alcohol.

Why:
Red-hot particles of fluffy chromic oxide are formed on ignition of ammonium dichromate. Some of the reaction product rolls over the sides of the reaction area and some shoots several feet in the air.

How:
Place the powdered chemical in a conical pile in the beaker. A roll of filter paper, about two inches in length, is soaked in the alcohol. This is inserted in the center of the cone. When lit, the wick burns rapidly, igniting the powder.

Suggestions:
In a darkened room the eruption is quite spectacular. Magnesium ribbon can be used as a fuse but is not as dependable as the alcohol-soaked wick. The experiment is not dangerous, but good ventilation is needed.

• Smoke Screens and Explosions

Action:
A few drops of a yellow solution when placed on a filter paper burst into flames and produce an intense white smoke. When the same solution is placed on a tablet of potassium chlorate, and is allowed to evaporate, a violent explosion occurs. Several candles ignite spontaneously when a drop of this liquid is placed upon them.

You Need:
White phosphorus; carbon disulfide; filter paper; potassium chlorate tablet; small candles; medicine dropper.

Why:
When exposed to the air, the carbon disulfide evaporates, leaving finely divided phosphorus, which ignites spontaneously, producing the white cloud, phosphorus pentoxide. Oxidation of phosphorus by chlorate is explosive.

How:
To prepare the yellow solution, cut a piece of white phosphorus under water, dry it, and dissolve it in the same volume of carbon disulfide.

Remarks:
The potassium chlorate tablet can be obtained from a chemical-

supply house. If, after 10 minutes, the tablet does not explode spontaneously, touch it with a meter stick. Take care, since the explosion is violent. Ordinary Christmas candles are best for the candle lighting. Roughen the wicks before applying the solution. There should be no tallow on the wicks. Allow 10–20 minutes for the candles to light.

CAUTION:
    The solution is very dangerous. Phosphorus burns are serious. Keep the solution in a glass-stoppered bottle; the solution should be further protected by placing the first bottle in a larger bottle.

VII. EXPLOSIONS

Chemical Cannon

Action:
    You drop a solid and liquid into a large test tube and quickly insert a stopper. Gas pressure will drive the stopper out of the tube with considerable force and with a loud pop.

You Need:
    Large test tube with cork; five grams sodium carbonate; 10 ml. vinegar.

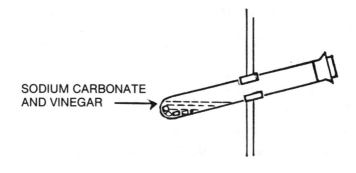

SODIUM CARBONATE
AND VINEGAR ———→

Why:

Carbonates and acid generate carbon dioxide gas, which when confined exerts pressure.

How:

Attach the test tube to a ring stand at a slight angle so that you or the audience will not be in the line of fire. A 200 mm. test tube is a good size.

Suggestions and CAUTION:

You can generate gases by carbonates and weak acids or by the action of active metals and hydrochloric acid. The cork must be fairly tight to get a loud pop. DO NOT STAND CLOSE TO THE TEST TUBE AS THE GAS IS GENERATED.

Dust Explosions

Action:

The demonstrator blows on a rubber tube fitted to the base of a large metal can. An explosion takes place that drives the cover of the can to the ceiling.

You Need:

One- or two-gallon can with tight-fitting cover; small metal funnel; 10–25 ml. dry lycopodium powder; rubber tube with clamp; candle.

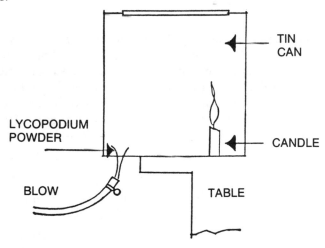

Why:
Explosions in flour mills and coal mines are simulated. The speed of oxidation of finely divided substances is illustrated in this combustion.

How:
In a hole in the base of the can place the upright funnel. Attach a rubber tube to the funnel into which you can blow. Place a clamp on the rubber tube near the funnel. When you are ready to demonstrate place a lighted candle in the base of the can, the lycopodium powder in the funnel and fasten the cover. As you blow in the tube, release the clamp.

Suggestions:
The candle can be placed in the can in a holder or suspended from the cover with a wire.

Use dried cornstarch or flour if you do not have lycopodium powder. The larger the can, the more violent the explosion. Holes in the sides of the can reduce the violence of the explosion.

Dust explosions can also be demonstrated by placing in a glass or paper tube some dried cornstarch or flour and blowing this powder into a burning candle or bunsen burner. Half a teaspoonful of powder can be blown with one puff. The huge flame is not expected.

Exploding Flame

Action:
From a rubber tube connected to the natural gas outlet you run gas into a hole on the top of the cover of a one-gallon syrup can. After three minutes the can will be filled with gas and you withdraw the tube. You light the gas as it escapes from the hole in the top of the can. The flame at first burns three or four inches high and then gradually subsides and seems to disappear after five minutes. Fifteen minutes later an explosion throws the cover into the air.

Why:
Explosions in combustible gases occur in air only when the ratio of gas to air reaches a critical value.

You Need:
One-gallon syrup or similar can with friction top cover. A hole one-eighth inch in diameter has been made in the center of the cover and in the center of the base.

Suggestions:
The explosion is not violent and therefore not dangerous. The experiment is a good one for a magic show since the explosion is unexpected after a lapse of fifteen minutes between the time that the flame disappears and the moment of the explosion.

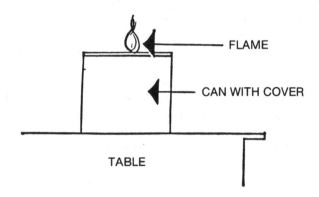

FLAME

CAN WITH COVER

TABLE

• Phosphorus Explosion

Action:
A vigorous explosion at the beginning of a chemical magic show puts the audience in the right mood for a series of mystery experiments. The explosion occurs back of the demonstration desk.

You Need:
Five-grain potassium chlorate tablet; iron rod; two grams yellow phosphorus; 10 ml. carbon disulfide; ring stand base.

Why:
Rapid oxidation of phosphorus in the presence of an oxidizing agent (chlorate) occurs with explosive violence.

How:

Previous to the performance of the experiment you place a few drops of a solution of yellow phosphorus in carbon disulfide on a five-grain tablet of potassium chlorate which is resting on the base of a ring stand. In fifteen minutes the solvent will have evaporated and the tablet is ready to be exploded. Touched with a metal rod from a ring stand, the explosion is violent.

CAUTION:

Cut yellow phosphorus under water. Handle with a large forceps. On dissolving the phosphorus in carbon disulfide you have a very dangerous solution since a drop of this material is highly combustible. You may keep it in a small dropper bottle, which is kept inside of a larger glass-stoppered wide-mouthed bottle.

This experiment must be performed only by someone fully aware of the dangers in handling phosphorus or the solution.

Potassium chlorate tablets can be purchased from a chemical distributor. Do not bring the hands or feet near the tablet when it is ready to explode. Under no condition allow the solution to come in contact with the skin or any flammable material.

• Nitrogen Triiodide Explosions

Action:

Walking into a room, a person is startled by sharp crackling sounds at his feet as he steps on small pieces of paper. Touching the paper lightly with a meter stick brings about small but sharp explosions.

You Need:

Five grams iodine; three grams potassium iodide; 20 ml. concentrated ammonium hydroxide; filter paper; funnel.

Why:

Nitrogen triiodide when dry explodes with the slightest disturbance.

How:

Stir the potassium iodide and iodine together in a beaker with 50 ml. of water. Add the ammonium hydroxide with stirring until no more precipitate forms. Filter and spread a thin layer of the wet solid on several filter papers. Break the filter papers into many small pieces and allow to dry for several hours.

On drying, the paper is extremely sensitive to touch and will explode violently with the slightest disturbance.

CAUTION:
The compound nitrogen triiodide can be safely handled when wet. Spreading a thin layer of the wet material on several pieces of filter paper lessens the violence of each explosion. It is not a powerful explosive, rather a sensitive one. The touch of a feather can set it off. UNDER NO CONDITION ALLOW ANY SIZEABLE QUANTITY OF THE DRY MATERIAL TO ACCUMULATE.

VIII. CRYSTALLIZATION AND PRECIPITATION

Chemical Garden

Action:
In a display case stands a large bottle nearly filled with a liquid. A small forest of trees appears to be growing in the liquid.

You Need:
Sodium silicate (water glass); large bottle or beaker; large crystals of salts such as cobalt chloride, ferric sulfate, nickel sulfate, manganous chloride, zinc sulfate and chromium nitrate.

Why:
A colloidal semipermeable membrane is formed around each salt. Water enters the sack diluting the concentrated solution. The sack breaks upward since there is greater pressure of water on the sides than from above. The growth therefore is upward. Colors formed are blue (cupric or cobalt), red-brown (ferric), green (nickel), white (zinc).

How:
Commercial sodium silicate (water glass) solution is well mixed with an equal quantity of water in a large bottle. Crystals are then

dropped so that they will be distributed on the bottom of the bottle. They begin to grow immediately. In a few hours some will have reached the top.

Suggestions:

In mixing with water, dilute the sodium silicate solution with water until the specific gravity is 1.1. Such a mixture may work better than one made up of equal quantities of water and commercial water glass solution.

If, after the chemical garden has been prepared and has been standing for some days, the solution is murky, you can clarify it by carefully syphoning off the solution and replacing the solution with water. The chemical garden is very fragile and should not be disturbed in the process.

Crystal Growth

Action:

On a demonstration table or display case is standing a large beaker filled with a liquid. Attached to a glass rod placed across the top of the beaker is a sheet of metal that extends into the solution. A growth of heavy shiny sheets of lead crystals hangs from the lower part of the metal.

You Need:

Two grams lead acetate in a large beaker of distilled water; a strip of zinc metal six inches long and an inch wide.

Why:

Displacement of lead ions by zinc results in a beautiful crystal display.

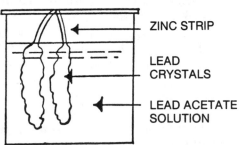

ZINC STRIP

LEAD CRYSTALS

LEAD ACETATE SOLUTION

Suggestions:

The solution of lead acetate, which may be slightly murky on mixing, will clear up on standing. Some of the zinc metal should be above the solution so that the viewer can see the change taking place. Although an immediate reaction occurs, crystals of appreciable size will not appear before 24 hours and they will then grow continuously. They may reach several inches in length, extending downward from the metal.

The murky solution of lead acetate can be cleared up by allowing it to stand for some hours and then filtering before setting up the demonstration.

Crystal Moss

Action:

In the display case stands a large flask containing a heavy growth of moss-like crystals. They almost fill the base of the flask.

You Need:

Large flask; narrow long coiled strips of zinc metal; five grams lead acetate dissolved in distilled water.

Why:

As zinc metal dissolves, small crystals of lead are precipitated. Oxidation of zinc metal is accompanied by reduction of lead ions.

How:

Cut sheet zinc into long, narrow strips. Coil them as you place them in the flask filled with the lead acetate solution. Suspend the zinc from a cork in the neck of the flask.

Suggestions:

Zinc reacts with the solution almost immediately, causing precipitation of moss-like lead crystals. Lead acetate solution is murky when first made but clears up in a day. This makes a beautiful chemical display after a day or two.

To make a clear solution of lead acetate, allow the solution to stand for some hours and then filter.

Crystals

Action:
  On display are several large crystals in a dish.

You Need:
  Ordinary alum (potassium aluminum sulfate).

Why:
  Crystals are seen to grow over a period of days.

How:
  Make a saturated solution of alum by heating the salt in a very small volume of water. Cool and allow crystals to settle. With a spatula or small knife remove some of the best crystals to a large dish. Again make a saturated solution of alum and pour this cool solution over the crystal. The crystal will grow and in a day or two should be removed and again placed in a fresh saturated solution. Repeat this procedure many times and eventually you will grow a large crystal.
  You must move the crystal around in many positions in order to get an even growth of faces.

Suggestions:
  Make a saturated solution of copper sulfate by heating 50 grams of salt in 100 ml. of water. Cool and suspend a crystal of copper sulfate in the solution. This crystal, fastened with a string to a ring stand, will continue to grow. After a day or two, remove the solution and replace it with a fresh saturated solution of copper sulfate. Repeating this procedure for days and weeks will eventually produce a giant crystal.

Lead Tree

Action:
  A tall beaker on display contains a beautiful lead tree growing downward.

You Need:
100 ml. water glass (specific gravity 1.06); 5 ml. saturated lead acetate solution; 6 ml. glacial acetic acid diluted to 100 ml.; paraffin; piece of mossy zinc; tall beaker.

Why:
This chemical display illustrates displacement of metals, speed of reaction and gel formation.

How:
The three solutions are mixed and stirred well in the beaker. If not acid to litmus, add more acid. Allow to stand. A gel is formed. After a day, push the piece of zinc into the surface of the gel and cover with melted paraffin. Slowly the crystals of lead proceed downward.

Suggestions:
You may wish to try a slight variation of the experiment. Place a piece of tin in the bottom of the tall beaker. Pour the mixed water glass and acid into the beaker. Let stand for a day. Add the saturated lead acetate to the surface. A lead tree will begin to form.

Orange Tree

Action:
An orange-colored crystalline tree is suspended as a decorative chemical display.

You Need:
200 grams potassium dichromate; package of pipe cleaners; 600 ml. beaker.

Why:
In a saturated solution orange-colored crystals grow.

How:
You prepare a saturated hot solution of potassium dichromate by placing 200 grams of the salt in a beaker and adding 200 ml. of water. Stir and heat to boiling.

You will now make a small fir tree with the pipe cleaners. Cut pieces, which will make up the branches. Fasten these to the stem with string. Then attach a string to the top of the tree, which is lowered into the beaker until it is completely covered, care being taken that it does not touch the sides. Support the arrangement from above and let it stand for a day or two. Crystals will grow on the stem and branches of the tree. Carefully lift the tree from the beaker and hang it up for a beautiful chemical display.

Use saturated solutions of other salts to make trees of other crystalline shapes and colors. You may try copper sulfate or colored alums.

Snow Tree

Action:
    A tree with crystals of snow is on display in a solution in a large beaker.

You Need:
    A thin copper sheet about seven inches long and five inches wide; a three-liter beaker filled with a solution of two grams of silver nitrate in distilled water.

Why:
    Copper is oxidized in a solution of silver nitrate. At the same time, silver ions are reduced to metallic silver.

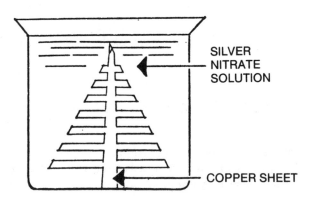

SILVER NITRATE SOLUTION

COPPER SHEET

How:
Cut the copper sheet into triangular shape and then into thin strips resembling branches of a tree. Suspend in the solution in the beaker. Copper metal goes into solution as beautiful silver crystals attach themselves to the branches. The crystals begin to grow immediately with the greatest concentration of crystals on the lower branches. In a few hours the tree will be quite complete. Color of the snow will range from white to rather gray in appearance.

Suggestions:
This is a good demonstration to perform at Christmas time. Cut right, the tree will look like a Christmas tree with silvery branches.

Freezing Without Cooling

Action:
A large Florence flask containing a clear colorless liquid is sitting on the demonstration table. You take a small crystal or two of a white salt and drop it into the flask. A beautiful star-shaped mass of white crystals immediately begins to radiate out from the point of impact, soon completely turning the liquid to a solid white mass. Turning the flask upside down, you show that all the liquid has frozen.

You Need:
Two- or 3-liter Florence flask; sodium acetate trihydrate.

Why:
The supersaturated solution of sodium acetate requires only one or two "seed" crystals of the salt to cause massive crystallization. So much solid is formed that the water is trapped in the mass and appears to be completely frozen.

How:
The supersaturated solution is prepared by heating in the flask sodium acetate trihydrate and water, in the ratio of 130 grams: 100 ml., until the salt has dissolved. Make up enough solution to

approximately half-fill the flask. Allow the solution to cool slowly without any disturbance.

Remarks:

Great care must be taken to avoid jarring the solution lest it crystallize too soon. Place a beaker over the mouth of the flask while it is cooling lest dust particles cause crystallization. This demonstration is a very striking example of the concept of supersaturation.

IX. FREEZING AND GEL FORMATION

• Cold Gas

Action:

You bubble natural gas through a long rubber tube that dips into a liquid in a 200 ml. beaker, which rests on a small wooden box. After a short period of bubbling you lift the beaker and show that it has frozen to the box. The box lifts using the beaker as a handle.

You Need:

100 ml. ether or concentrated ammonium hydroxide in the beaker; source of natural gas; small wooden box; rubber tubing.

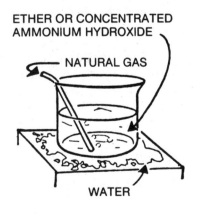

ETHER OR CONCENTRATED
AMMONIUM HYDROXIDE

NATURAL GAS

WATER

Why:
Rapid loss of heat in evaporation of ether or ammonia lowers the temperature of water below its freezing point.

How:
Wet the surface of the box with water. Place the beaker containing the ether or ammonium hydroxide on the wet surface. Bubble natural gas into the liquid.

CAUTION:
Ether and natural gas are very flammable. No flames should be present. Ammonia fumes make the demonstration impractical in a small room.

Fast Freezing

Action:
You place a white powder in a 400 ml. beaker. Put the beaker on the wet top of a small inverted wooden box. With rapid stirring, pour 100 ml. of water into the beaker. In a minute or two, you lift the beaker and the box comes along with it since it is frozen fast to the beaker.

You Need:
100 grams ammonium nitrate; small wooden box such as a chalk box; thermometer reading at least ten degrees below the freezing point of water.

Why:
A salt, which absorbs heat on dissolving, lowers the temperature of its solution below the freezing point of water.

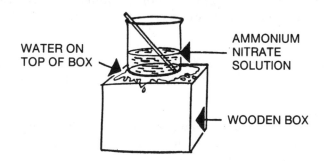

WATER ON
TOP OF BOX

AMMONIUM
NITRATE
SOLUTION

WOODEN BOX

How:

On dropping the powder into the water and stirring rapidly, heat is taken out of the solution and the temperature drops rapidly. The bottom of the beaker is cooled below the freezing point of water. The water below the beaker is frozen. This binds the beaker to the box.

Suggestions:

Do not spill any of the salt on the wet top of the box. This will prevent formation of ice where the beaker and box come in contact.

Recording the temperature at intervals makes this an interesting class project.

The beaker and ice combination can be passed around the class to show the interesting ice formation.

Hard Water

Action:

Two beakers containing colorless liquids are held in the hands. The two are poured back and forth and suddenly the liquid solidifies. When the beaker is inverted, no liquid runs out.

Some of the solid is removed with a spatula and lit with a match. It burns with a hot flame.

You Need:

300 ml. ethyl alcohol; saturated solution of calcium acetate made by stirring 12 grams of the solid in 40 ml. of water; two 400 ml. beakers; spatula.

Why:

A colloidal substance, solid alcohol or "canned heat," is formed by mixing calcium acetate solution and ethyl alcohol.

How:

One beaker contains the alcohol, the other saturated calcium acetate solution. Held high for all to observe, the liquids are mixed.

Suggestions:

If the experiment does not work it is probably because you do not have a saturated calcium acetate solution. It takes a large amount of the salt to make a saturated solution.

The flame is blue and almost invisible. To emphasize flammability of the solid you may sprinkle a little lithium salt on the flame.

If you wish to emphasize color of the alcohol add a few drops of food coloring. This makes the experiment easier to follow.

To make the liquids seem to disappear, perform the experiment in stainless steel beakers. On mixing and inverting, the material seems to have vanished.

Regelation

Action:
Heavy weights are attached to each end of a wire. The wire is stretched over a chunk of ice. Slowly the ice is cut into two pieces, but examination of the ice afterward shows that it is still intact.

Why:
The pressure of the wire on the ice decreases the melting point, causing local melting. As the wire passes through, the water refreezes above it.

X. SMOKE AND VAPORS

• Violet Vapors

Action:
A stream of water sprayed into an evaporating dish gives rise to violet-colored vapors.

You Need:
One part powdered zinc to four parts powdered iodine, an evaporating dish, a wash bottle or medicine dropper.

Why:
Water provides the vehicle for reaction between zinc and iodine.

How:

Powder the zinc and iodine separately in a mortar and pestle, then mix well. When water strikes the mixture an immediate chemical reaction follows with violet vapors of iodine being given off. Holding a white card behind the dish gives emphasis to the color.

CAUTION:

Solid or vaporous iodine is very corrosive and causes severe burns.

• Violet Smoke

Action:

A stream of water sprayed into an evaporating dish gives rise to copious clouds of violet-colored smoke.

You Need:

Four grams powdered zinc; four grams powdered ammonium nitrate; half gram iodine crystals; an evaporating dish; a wash bottle or medicine dropper.

Why:

Water initiates the reaction. Zinc oxide particles mixed with violet vapors of iodine create colored smoke.

How:

Powder the ingredients separately. Mix well and place in the evaporating dish. A small stream of water from a wash bottle or medicine dropper will start the reaction. The smoke is white zinc oxide intermingled with iodine that has sublimed in the heat.

Suggestions:

If the room is small reduce the quantity of chemicals since the smoke is quite heavy. In a large auditorium the smoke will probably not cause any discomfort. You may wish to perform this experiment shortly before the close of the chemistry magic performance.

A white background helps to bring out the violet coloration.

To make white smoke merely leave out the iodine.

CAUTION:
Iodine, solid or vaporous, is corrosive.

• Smoke

Action:
A few drops of liquid from a medicine dropper fall in a heated crucible. Dense white smoke instantly emerges.

You Need:
Equal quantities of powdered silica and zinc; carbon tetrachloride.

Why:
Smoke probably consists of solid particles of zinc oxide in the presence of several products.

How:
After mixing the powders, place them in a crucible and heat for two or three minutes. From a medicine dropper you add carbon tetrachloride a few drops at a time. The reaction is instantaneous.

CAUTION:
The demonstration produces obnoxious and choking fumes so you may wish to perform the experiment in a fume hood or out-of-doors.

• Smoke Blown Through Glass

Action:
Two empty tumblers stand on the table some distance apart. You tip them upside down to show that they are empty. Now place one mouth to mouth over the other at the same time as you cover them with a cloth. Standing to one side you blow cigarette smoke at the glasses, remove the cloth and they are full of smoke.

You Need:
A film of concentrated hydrochloric acid in one tumbler and concentrated ammonium hydroxide in the other.

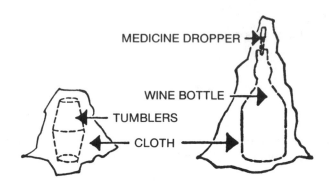

Why:

Solid white smoke particles of ammonium chloride are produced when the acid fumes are in contact with ammonia fumes.

How:

A few drops of the reagents are needed in the tumblers. Only enough to form a thin film is necessary. The cloth must be placed over the tumblers at the same instant as one is inverted over the other since the chemicals react to form ammonium chloride almost immediately.

Suggestions:

An interesting variation of this experiment is to blow smoke into a large transparent wine bottle. Before the demonstration pour a few drops of the acid into the bottle. Invert the bottle and form a good film throughout. Place a few drops of ammonia in the bulb of a medicine dropper that you have concealed in a large cloth. As you bring the cloth over the bottle, squeeze the bulb so that a few drops of ammonia enter the wine bottle. Stand to one side and blow cigarette smoke into the bottle. When you remove the cloth, there will be a heavy concentration of smoke in the bottle.

CAUTION:

Be careful when working with concentrated acid and ammonia.

• Smoke Producer

Action:
At the will of the demonstrator, white smoke shoots out of a tube.

You Need:
Two bottles with 2-hole rubber stoppers; rubber bulb and rubber tubing; concentrated hydrochloric acid; concentrated ammonium hydroxide.

Why:
Ammonium chloride particles make up the white smoke. Smoke is formed when fumes of hydrogen chloride come in contact with ammonia.

How:
Place a few ml. of the acid in the first bottle and the same amount of the ammonia in the second bottle. Arrange the apparatus so that pressure on the bulb causes air to pass first to the bottle containing the acid. A glass tube extends from the acid bottle to the bottom of the ammonia bottle. Dense fumes generated in the second bottle are then conducted out of the ammonia bottle in a rubber tube.
To generate smoke, merely press on the rubber bulb.

Suggestions:
To make a steady stream of smoke remove the rubber bulb. Use your breath to blow into the first bottle.

CAUTION:
Do not inhale the acid! Hydrochloric acid is very corrosive.

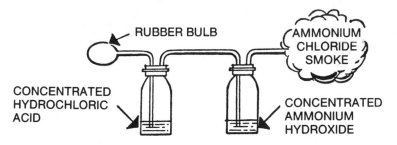

RUBBER BULB

AMMONIUM CHLORIDE SMOKE

CONCENTRATED HYDROCHLORIC ACID

CONCENTRATED AMMONIUM HYDROXIDE

• Smoke Rings

Action:

White smoke rings rise from an intermittent flame that bursts above water in a transparent dish.

You Need:

500 ml. distilling flask; 200 ml. 40% potassium hydroxide solution; three or four small pieces of phosphorus; a large beaker.

Why:

Bubbles of phosphine in the presence of P_2H_2 escape from the tube in the beaker. On contact with air they ignite. The exploding bubbles form smoke rings in the form of halos that rise in the room.

How:

Attach the flask to a ring stand. Add the phosphorus and the base solution. Connect a rubber tube to the condensate tube of the flask. To the end of this attach glass tubing. This terminates under water in the beaker just below the surface. The open end extends upward.

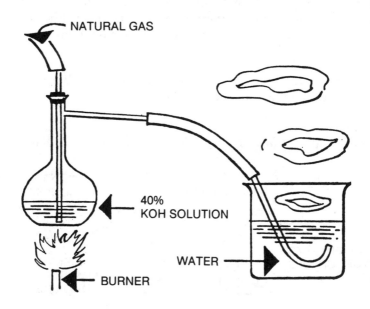

NATURAL GAS

40%
KOH SOLUTION

WATER

BURNER

Fit a one-hole stopper in the flask. A glass tube extends through it into the solution, the other end of the tube being attached to a rubber hose connected to the gas supply.

Allow the gas to bubble through the solution and then heat it to boiling. Smoke rings will then begin to form.

To stop the flow of gas, remove the flame under the flask and pour water into the beaker. Water will be drawn into the flask as the steam cools. This prevents fire in the flask when disassembling the apparatus.

Suggestions and CAUTIONS:

Unless you are familiar with the flammable properties of phosphorus do not attempt this experiment. Phosphorus must be cut under water and handled with forceps.

If you do not have a distilling flask, you may use an ordinary flask with a two-hole rubber stopper.

Smoke rings can also be made by the use of a few lumps of calcium phosphide. Drop the chemical into a glass cylinder. Bubbles of phosphine form and explode in contact with air, forming smoke rings of phosphorus pentoxide. If you invert a glass funnel over the calcium phosphide, you will tend to concentrate the escaping bubbles so that they will escape in the center of the cylinder. *Be careful!* Phosphine is very poisonous.

XI. SPECIFIC GRAVITY

• Floaters

I

Action:

On display is a 500 ml. graduate filled with liquids at four levels. Solid objects float at each level.

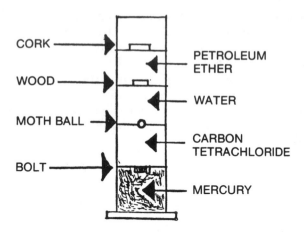

You Need:
100 ml. each of mercury, carbon tetrachloride, water and gasoline or petroleum ether; small screws or bolts; moth ball; small wood block and cork.

Why:
Each solid is buoyed up by the weight of the liquid displaced.

How:
Pour the mercury into the graduate and drop a couple of screws or bolts on the surface. Add the carbon tetrachloride and drop the moth ball on its surface. Water is now added and the wood piece is placed on it. The upper liquid is petroleum ether or gasoline on which cork is floating.

The wood piece may rise into ether unless weighted with metal. To do this, bore holes in the wood and insert screws until the weight is great enough to float on water but sink in ether or gasoline.

Suggestions:
Carbon tetrachloride and ether can be replaced by other liquids of equal density but they must be immiscible with water.

The demonstration can be made especially attractive by placing on each side of the display similar graduates with colored liquids.

A tight-fitting cork in the graduate prevents evaporation of the highly volatile ether.

CAUTIONS:
All of the liquids except water are potentially dangerous and should be handled carefully. Ventilation should be good, and no flames should be present.

II

An interesting, but less permanent, variation is the following:

Action:
A tall bottle is exhibited containing several colored liquids in layers. At the bottom is a silver liquid. Above that, in order: a white liquid, a blue liquid, a colorless liquid and another blue liquid. On each line of demarcation of these liquids, several balls can be seen suspended in the bottle.

You Need:
Mercury for the silver liquid; saturated solution of zinc sulfate for the white; solution of copper sulfate for the blue; water for the colorless; and methylene blue in alcohol solution for the top blue.

Why:
The balls are made of iron, which floats on mercury; wax mixed with shot, which floats on zinc sulfate solution, but not on copper sulfate; wax with less shot, which floats on copper sulfate but not on water; and cork, which floats on water, but not on alcohol.

Suggestion:
Care must be taken that the different liquids are not mixed when added to the bottle. The liquids are poured into the bottle in turn with a funnel and a glass tube. A visible line of demarcation is clear for a considerable time because of the slow rate of diffusion of the solutions.

Wine Tower

Action:

On a low plate are standing two drinking glasses full of water. One glass is inverted over the other. On the base of the upper inverted glass stands a full wine glass upright. Two wicks are hanging over the side of the wine glass. Wine is dripping from the wicks to the outside of the upper glass. The dripping liquid gradually enters the inverted glass at its junction with the upright glass. The red wine gradually replaces the colorless water in the upper inverted drinking glass.

You Need:

Two drinking glasses, wine glass, two lamp wicks, low plate, 200 ml. ethyl alcohol colored with red food coloring.

Why:

Capillary action draws wine over the wine glass. The colored liquid then drips down the sides of the upper glass. Where the glasses touch, the wine enters, curves upward and rises to the top of the inverted glass. Water in the upper glass is being displaced by the lighter colored alcohol.

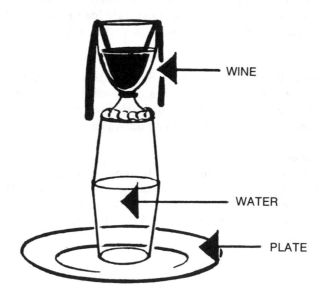

How:

On a plate place a full glass of water. Invert a similar full glass of water over it, mouth to mouth. Use a sheet of paper held firmly over the top as you invert one glass over the other.

Hang two wicks over the edge of the wine glass and fill this glass with the alcohol colored with a little red food coloring. Place the wine glass on the base of the upper inverted drinking glass. Arrange the wicks so that the dripping liquid falls on the outside surface of the upper inverted glass.

Suggestions:

The gradual displacement of the water by "wine" in the inverted tumbler is fascinating to watch. An hour or two may elapse before there will be much displacement of water by the "wine." This demonstration can very well be done in a display case. Wine with high alcoholic content and dark color can be used in the wine glass in place of the colored alcohol.

XII. POLYMERIZATION

Synthetic Rubber

Action:

Two beakers stand on the demonstration table. One contains a clear liquid; the other a milky-appearing substance. You take these beakers in your hands, pour their contents into a large beaker and stir with a glass rod. A rubbery solid appears to coagulate. With the hands the liquid is squeezed out of the rubber ball. You strike the floor with the ball and it rebounds to the ceiling. Now you pass the ball around the room for inspection.

You Need:
A small beaker half full of synthetic rubber latex; a similar beaker with 10% acetic acid.

Why:
The latex is coagulated by acid. The product has properties of rubber.

How:
One solution is 10% acetic acid; the other is synthetic rubber latex.

Suggestions:
The solutions must be well mixed with a spatula or stirring rod to get the maximum amount of coagulation. The mixture does not harm the hands and the liquid can easily be squeezed out of the rubber ball.

Do not spill the latex on your clothes since it is almost impossible to remove.

With a quart of synthetic latex you can do the experiment fifty or more times.

Nylon Rope Trick

Action:
Two beakers, each containing a clear liquid, are resting on the demonstration table. You pick up each beaker, hold one at a tilted angle, and slowly pour the contents of the other into it so that the liquid runs down the side of the beaker. Using metal tweezers, you grab the membrane which has formed at the interface of the liquids and slowly pull up. A strand of nylon rope continues to issue from the beaker indefinitely.

You Need:
Two solutions, A and B, made from the following ingredients.
A: 6 grams sebacoyl chloride; 70 ml. heptane.
B: 3 grams of 1,6–diaminohexane; 70 ml. water.
Metal tweezers; two 150 ml. beakers; glass rod.

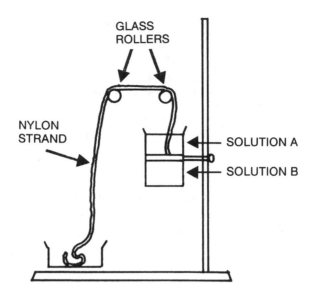

Why:
   The acyl chloride of any dicarboxylic acid will polymerize with
any diamine by substitution reaction to produce a nylon polymer.
HCl is the other product.

How:
   Solutions A and B are prepared by simply dissolving the acyl
chloride in heptane and the diamine in water. Use equal volumes of
each solution for the demonstration. Slightly tilt the beaker
containing solution B and carefully pour solution A down the side
of it so that two layers form. Use a glass rod to wrap strands of two
feet or so of the nylon around the tweezers as you pull it out. Later,
put the mass of nylon in 500 ml. methanol to remove the acid that
accompanies it, thus preserving it. You can make a ball of nylon
also.

Remarks:
   The demonstration can be improved to show the dramatic
production of nylon by fashioning a continuous rope of polymer
that issues indefinitely from the beaker. Clamp the beaker to a ring
stand over which is fixed a set of two rollers made from glass
tubing. By pulling the nylon up and over the rollers, you can cause
the strand to run into a container set below.

XIII. DELAYED OR CONSECUTIVE REACTIONS

• Disappearing Blue

Action:
A rubber-stoppered one-liter flask containing a colorless liquid is standing on the demonstration table. You pick it up, give it one quick jerk upward and the solution turns dark blue. On standing for ten seconds, the color changes to pink and then to colorless again.

You Need:
Five grams potassium hydroxide; three grams dextrose; pinch of methylene blue; one-liter rubber-stoppered flask.

Why:
Action of oxygen in the air on methylene blue causes a color change by oxidation.

How:
Dissolve the ingredients in 250 ml. of water and place in the flask.

Suggestions:
One quick jerk upward is all that you need to make a color change. Violent shaking makes the blue color persist for a greater number of seconds and detracts from the demonstration.

When once prepared the experiment can be repeated many times. After two or three days the chemicals seem to lose their effectiveness.

Keep the amount of methylene blue very small.

CAUTION:
Be careful when working with potassium hydroxide.

Disappearing Orangeade

Action:

Two colorless solutions in beakers stand on the table. You pour liquid from the first beaker into the second. An orange color appears and then disappears where the two solutions meet.

Pouring from the second beaker into the first produces a bright orange color. On addition of all the solution the color completely disappears.

You Need:

Five grams mercuric chloride; five grams potassium iodide; distilled water; two 400 ml. beakers.

Why:

Mercuric chloride reacts with potassium iodide to produce orange mercuric iodide. This substance in turn dissolves in potassium iodide, producing the colorless complex of potassium iodide and mercuric iodide.

How:

The first beaker is two thirds full of dilute mercuric chloride solution. This is made by dissolving the salt in 300 ml. of distilled water. The second beaker is half filled with dilute potassium iodide solution. Dissolve the salt in 200 ml. of water to make this solution.

A variation of this demonstration is one that can be called "Halloween Colors." You have three beakers containing equal volumes of colorless solutions on the demonstration table before you. You have labeled these beakers A, B and C. You pour B into C and this mixture into A, which you hold high for all to see the remarkable color change that occurs in a few seconds. The colorless solution suddenly becomes bright orange. In a second or two the orange color changes to jet black. Solutions are made as follows:

A. Fifteen grams potassium iodate in 1000 ml. water.

B. Four grams soluble starch in 500 ml. boiling water. Fifteen grams of sodium bisulfite are dissolved in 500 ml. water. Mix the two solutions to make 1000 ml. of solution B.

C. Three grams mercuric chloride in 1000 ml. water.

Magic Ink

Action:
You hold two beakers before the audience. The beakers, containing colorless solutions, are mixed and after a few moments you make a pass with the hand, causing ink to form.

You Need:
Two solutions that we will call A and B are prepared. Solution A is made by dissolving 0.5 gram potassium iodate in water, to make a total volume of 300 ml. Solution B is made by mixing 0.2 gram sodium sulfite in a few ml. of water with 1 ml. dilute sulfuric acid and adding to this 15 ml. of stable starch solution. Solution B is likewise made up to 300 ml.

Why:
After you mix the solutions, a period of time elapses before the reaction occurs, liberating iodine, which in turn colors the starch.

Suggestions:
By practice, you can predict the exact number of seconds that will elapse before the color change. A few magic words at this time and a pass or two of the hand over the mixed solutions will make the experiment an amusing one.

Slight variations in concentrations as well as temperature will effect the time lapse between the time of mixing and the color change.

If you do not have a stable starch solution, it is possible to make the solution. Dissolve two grams of ordinary starch in 100 ml. of water, heat to boiling, filter and use this in diluting solution B.

Synthetic Gold

Action:
Two colorless solutions in beakers are standing on the demonstration table. You pour one solution into the other. At first

nothing happens, but in about thirty seconds a beautiful gold color appears.

You Need:
Two solutions called here A and B. You can make them up in the quantities indicated for one performance or in greater quantity if you wish to repeat the experiment more than once.

Solution A is made by stirring together 1 gram sodium arsenite in 50 ml. water and then mixing 5.5 ml. glacial acetic acid with the resulting solution.

Solution B is made by stirring 10 grams photographer's "hypo" (sodium thiosulfate) in 50 ml. water.

Why:
Delay in the color change is probably due to the slow reaction between the acid and "hypo" that results in liberation of hydrogen sulfide gas. Its reaction in turn with sodium arsenite results in the precipitation of yellow arsenious sulfide.

How:
Practice the experiment with a stop watch so that you can predict the exact number of seconds that will elapse between the time of mixing and color appearance. You may wish to tell a story that reaches its climax at the moment of color change. Use a magic wand, make a pass with the hand or use magic words at the appropriate time.

• Oscillating Clock Reaction

Action:
On the demonstration table are four beakers containing clear liquids and a large graduated cylinder. You add, successively, equal volumes of three of the liquids to the cylinder and stir, whereupon the mixture changes from colorless to yellow. This change oscillates back and forth in color. Then you add a little of the fourth liquid, whereupon the oscillating color changes are colorless to yellow to blue. The process continues for at least 15 minutes.

You Need:
Four solutions, Λ, B, C and D, made from the following ingredients.

A: 3 ml. concentrated sulfuric acid; 500 ml. water; 21 grams potassium iodate.

B: 180 ml. 30% hydrogen peroxide; 320 ml. water.

C: 8 grams malonic acid; 1.9 grams manganese (II) sulfate dihydrate; 500 ml. water.

D: 2.5 grams potato starch; 500 ml. water.

500 ml. graduated cylinder; glass stirrer.

Why:
The yellowish color is due to formation of iodine by reduction of iodate. The blue is caused by complexing of iodine with starch. The reaction mechanism has not yet been determined.

How:
Solution A is prepared by adding the sulfuric acid to the water. Stir and add the potassium iodate. Solutions B and C are made by simply adding together the ingredients. Solution D is made by first making a thin paste of starch with water and dribbling it into 500 ml. of boiling water. Let the mixture boil for 2–3 minutes and then cool. Better results may be obtained by using more starch.

Use equal volumes of solutions A, B and C at first. Then add 2 ml. of solution D.

Remarks:
Use of a lighted magnetic stirrer makes the demonstration more dramatic.

CAUTION:
Be careful when handling concentrated sulfuric acid.

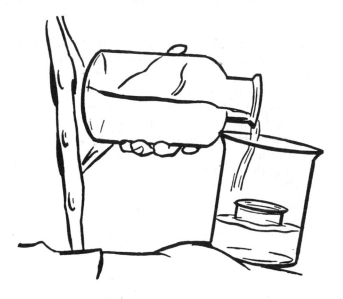

XIV. MISCELLANEOUS

Disappearing Beaker

Action:
You hold a large and a small beaker before the audience. You place the small beaker in the large one. Then you pour a colorless liquid from a curiously shaped bottle into the large beaker, causing the small beaker to completely disappear in the liquid.

You Need:
590 ml. carbon tetrachloride; 410 ml. benzene; 2-liter and 100-ml. beakers; large bottle.

Why:
The liquid has the same refractive index as pyrex glass. Therefore, the small pyrex beaker becomes invisible when surrounded by the liquid.

How:
To prepare the liquid you merely mix the carbon tetrachloride and benzene in the proportions given. Then pour this liquid into the bottle.

Suggestions:

You can perform another entertaining demonstration with the material mentioned above. In addition you will need a short piece of pyrex rod or tubing. Before the performance you place the small beaker in the large one and pour in the solution. The small beaker becomes invisible. You are now ready for the performance.

With one hand you slowly lower the glass rod into the solution. It disappears as it enters the solution until none of it is visible. You drop it. With the other hand remove the beaker, giving the impression that the glass rod was transformed into a beaker.

Placing a broken beaker into the solution and withdrawing a whole one makes another interesting variation.

The liquid has about the same index of refraction as an ordinary drinking glass. With it you can cause a drinking glass to disappear.

Osmosis

Action:

A carrot, suspended in a beaker and to which is attached an upright piece of glass tubing, is standing on the demonstration table. Water has risen in the tube to a height of two feet.

You Need:

Carrot; sugar; beaker; small one-hole rubber stopper attached to three feet of glass tubing.

Why:

Water in the beaker moves into the carrot by osmosis. The sugar solution becomes diluted and moves upward in the tube, exerting a pressure, called osmotic pressure.

How:

Using a cork borer make a hole in the top center of a large carrot. The hole should be deep enough to hold 10 or 15 ml. of a saturated sugar solution. It should be of a size to permit a tight fit of a rubber stopper. Attach the assembly to a ring stand and place in a beaker of water.

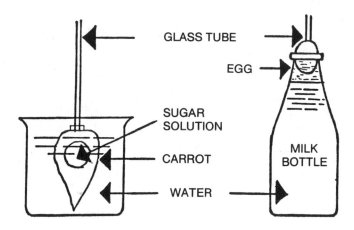

Suggestions:
The movement of water upward in the tube takes many hours. The experiment is not spectacular but is interesting as a class project. If the experiment does not go well, it may be that you have a leak between the stopper and the carrot. A potato or parsnip may be substituted in place of the carrot.

The following simple demonstration of osmosis is easy to perform. Tap one end of a fresh egg lightly. Remove some of the shell but do not damage the membrane. Place this end of the egg in the top of a milk bottle filled with water. Make a small hole in the upper end of the egg. Insert a six-inch piece of glass tubing. Seal with candle drippings.

In half an hour water will rise in the glass tube and continue to do so for several hours. The water will rise to a height of several inches.

• Pulsing Mercury

Action:
A small amount of mercury metal, covered with a colorless liquid, is contained in a shallow watch glass. An iron needle embedded in a wooden handle is fastened so that it touches the mercury. The mercury oscillates constantly, moving toward the iron, then away from it.

You Need:
Mercury metal; iron or steel needle; wooden dowel; clamp; watch glass; dilute sulfuric acid.

Why:
The liquid, which is dilute sulfuric acid, preferentially attacks the iron rather than the mercury, setting up an electric cell. Oxidation of the iron builds up a negative charge on the needle, while reduction at the mercury surface builds up a positive charge. Thus the mercury attracts to the needle and discharge of electricity occurs. Negative charge repulsion causes the mercury to contract, etc.

Remarks:
This demonstration is best adapted to small groups and can be set up with a minimum of equipment. It illustrates the principles of electrochemistry and the electrochemical series of metals.

CAUTION:
Be careful when using mercury and sulfuric acid.

• Glass Dissolves in Water

Action:
A clean test tube is broken and ground up in a clean mortar. When the ground glass is placed in a large test tube almost filled with water the solution becomes faint pink.

You Need:
Small soft glass test tube; large test tube; mortar and pestle; phenolphthalein solution.

Why:
Finely divided glass slowly dissolves to a limited extent in water. Hydrolysis occurs producing silicic acid and a weakly alkaline solution. The basic solution turns the indicator pink.

How:
Place about 10 drops of phenolphthalein solution in the water.

Remarks:
It probably will take some time for the glass to dissolve and the pink to form. It is necessary to clean all apparatus thoroughly before the demonstration.

CAUTION:
Be careful when grinding the test tube. Inhaling glass dust can be dangerous.

Vibratory Motion

Action:
Two pendulums hang suspended two or three feet apart. You start one of them swinging. After a short time, the second pendulum begins to swing. The first stops, but the second keeps swinging. When this one stops, the other begins again, and they alternate for a considerable time.

Why:
Each pendulum in motion sets up sympathetic vibrations in the other. The loss of energy from the first to the second stops the first.

How:
The bobs are suspended from a horizontal string by wires of equal length.

Remarks:
The demonstration is impressive because the motion continues for a long time, and is clearly visible to an audience of any size. It is an effective illustration of the principle of simple harmonic motion.

A CATALOG OF SELECTED
DOVER BOOKS
IN ALL FIELDS OF INTEREST

A CATALOG OF SELECTED DOVER
BOOKS IN ALL FIELDS OF INTEREST

CONCERNING THE SPIRITUAL IN ART, Wassily Kandinsky. Pioneering work by father of abstract art. Thoughts on color theory, nature of art. Analysis of earlier masters. 12 illustrations. 80pp. of text. 5⅜ x 8½. 0-486-23411-8

CELTIC ART: The Methods of Construction, George Bain. Simple geometric techniques for making Celtic interlacements, spirals, Kells-type initials, animals, humans, etc. Over 500 illustrations. 160pp. 9 x 12. (Available in U.S. only.) 0-486-22923-8

AN ATLAS OF ANATOMY FOR ARTISTS, Fritz Schider. Most thorough reference work on art anatomy in the world. Hundreds of illustrations, including selections from works by Vesalius, Leonardo, Goya, Ingres, Michelangelo, others. 593 illustrations. 192pp. 7⅛ x 10¼. 0-486-20241-0

CELTIC HAND STROKE-BY-STROKE (Irish Half-Uncial from "The Book of Kells"): An Arthur Baker Calligraphy Manual, Arthur Baker. Complete guide to creating each letter of the alphabet in distinctive Celtic manner. Covers hand position, strokes, pens, inks, paper, more. Illustrated. 48pp. 8¼ x 11. 0-486-24336-2

EASY ORIGAMI, John Montroll. Charming collection of 32 projects (hat, cup, pelican, piano, swan, many more) specially designed for the novice origami hobbyist. Clearly illustrated easy-to-follow instructions insure that even beginning papercrafters will achieve successful results. 48pp. 8¼ x 11. 0-486-27298-2

BLOOMINGDALE'S ILLUSTRATED 1886 CATALOG: Fashions, Dry Goods and Housewares, Bloomingdale Brothers. Famed merchants' extremely rare catalog depicting about 1,700 products: clothing, housewares, firearms, dry goods, jewelry, more. Invaluable for dating, identifying vintage items. Also, copyright-free graphics for artists, designers. Co-published with Henry Ford Museum & Greenfield Village. 160pp. 8¼ x 11. 0-486-25780-0

THE ART OF WORLDLY WISDOM, Baltasar Gracian. "Think with the few and speak with the many," "Friends are a second existence," and "Be able to forget" are among this 1637 volume's 300 pithy maxims. A perfect source of mental and spiritual refreshment, it can be opened at random and appreciated either in brief or at length. 128pp. 5⅜ x 8½. 0-486-44034-6

JOHNSON'S DICTIONARY: A Modern Selection, Samuel Johnson (E. L. McAdam and George Milne, eds.). This modern version reduces the original 1755 edition's 2,300 pages of definitions and literary examples to a more manageable length, retaining the verbal pleasure and historical curiosity of the original. 480pp. 5⁵⁄₁₆ x 8¼. 0-486-44089-3

ADVENTURES OF HUCKLEBERRY FINN, Mark Twain, Illustrated by E. W. Kemble. A work of eternal richness and complexity, a source of ongoing critical debate, and a literary landmark, Twain's 1885 masterpiece about a barefoot boy's journey of self-discovery has enthralled readers around the world. This handsome clothbound reproduction of the first edition features all 174 of the original black-and-white illustrations. 368pp. 5⅜ x 8½. 0-486-44322-1

STICKLEY CRAFTSMAN FURNITURE CATALOGS, Gustav Stickley and L. & J. G. Stickley. Beautiful, functional furniture in two authentic catalogs from 1910. 594 illustrations, including 277 photos, show settles, rockers, armchairs, reclining chairs, bookcases, desks, tables. 183pp. 6½ x 9¼. 0-486-23838-5

AMERICAN LOCOMOTIVES IN HISTORIC PHOTOGRAPHS: 1858 to 1949, Ron Ziel (ed.). A rare collection of 126 meticulously detailed official photographs, called "builder portraits," of American locomotives that majestically chronicle the rise of steam locomotive power in America. Introduction. Detailed captions. xi+129pp. 9 x 12. 0-486-27393-8

AMERICA'S LIGHTHOUSES: An Illustrated History, Francis Ross Holland, Jr. Delightfully written, profusely illustrated fact-filled survey of over 200 American lighthouses since 1716. History, anecdotes, technological advances, more. 240pp. 8 x 10¾. 0-486-25576-X

TOWARDS A NEW ARCHITECTURE, Le Corbusier. Pioneering manifesto by founder of "International School." Technical and aesthetic theories, views of industry, economics, relation of form to function, "mass-production split" and much more. Profusely illustrated. 320pp. 6⅛ x 9¼. (Available in U.S. only.) 0-486-25023-7

HOW THE OTHER HALF LIVES, Jacob Riis. Famous journalistic record, exposing poverty and degradation of New York slums around 1900, by major social reformer. 100 striking and influential photographs. 233pp. 10 x 7⅞. 0-486-22012-5

FRUIT KEY AND TWIG KEY TO TREES AND SHRUBS, William M. Harlow. One of the handiest and most widely used identification aids. Fruit key covers 120 deciduous and evergreen species; twig key 160 deciduous species. Easily used. Over 300 photographs. 126pp. 5⅜ x 8½. 0-486-20511-8

COMMON BIRD SONGS, Dr. Donald J. Borror. Songs of 60 most common U.S. birds: robins, sparrows, cardinals, bluejays, finches, more—arranged in order of increasing complexity. Up to 9 variations of songs of each species.
Cassette and manual 0-486-99911-4

ORCHIDS AS HOUSE PLANTS, Rebecca Tyson Northen. Grow cattleyas and many other kinds of orchids—in a window, in a case, or under artificial light. 63 illustrations. 148pp. 5⅜ x 8½. 0-486-23261-1

MONSTER MAZES, Dave Phillips. Masterful mazes at four levels of difficulty. Avoid deadly perils and evil creatures to find magical treasures. Solutions for all 32 exciting illustrated puzzles. 48pp. 8¼ x 11. 0-486-26005-4

MOZART'S DON GIOVANNI (DOVER OPERA LIBRETTO SERIES), Wolfgang Amadeus Mozart. Introduced and translated by Ellen H. Bleiler. Standard Italian libretto, with complete English translation. Convenient and thoroughly portable—an ideal companion for reading along with a recording or the performance itself. Introduction. List of characters. Plot summary. 121pp. 5¼ x 8½. 0-486-24944-1

FRANK LLOYD WRIGHT'S DANA HOUSE, Donald Hoffmann. Pictorial essay of residential masterpiece with over 160 interior and exterior photos, plans, elevations, sketches and studies. 128pp. 9¼ x 10¾. 0-486-29120-0

THE CLARINET AND CLARINET PLAYING, David Pino. Lively, comprehensive work features suggestions about technique, musicianship, and musical interpretation, as well as guidelines for teaching, making your own reeds, and preparing for public performance. Includes an intriguing look at clarinet history. "A godsend," *The Clarinet,* Journal of the International Clarinet Society. Appendixes. 7 illus. 320pp. 5⅜ x 8½. 0-486-40270-3

HOLLYWOOD GLAMOR PORTRAITS, John Kobal (ed.). 145 photos from 1926-49. Harlow, Gable, Bogart, Bacall; 94 stars in all. Full background on photographers, technical aspects. 160pp. 8⅜ x 11¼. 0-486-23352-9

THE RAVEN AND OTHER FAVORITE POEMS, Edgar Allan Poe. Over 40 of the author's most memorable poems: "The Bells," "Ulalume," "Israfel," "To Helen," "The Conqueror Worm," "Eldorado," "Annabel Lee," many more. Alphabetic lists of titles and first lines. 64pp. 5³⁄₁₆ x 8¼. 0-486-26685-0

PERSONAL MEMOIRS OF U. S. GRANT, Ulysses Simpson Grant. Intelligent, deeply moving firsthand account of Civil War campaigns, considered by many the finest military memoirs ever written. Includes letters, historic photographs, maps and more. 528pp. 6⅛ x 9¼. 0-486-28587-1

ANCIENT EGYPTIAN MATERIALS AND INDUSTRIES, A. Lucas and J. Harris. Fascinating, comprehensive, thoroughly documented text describes this ancient civilization's vast resources and the processes that incorporated them in daily life, including the use of animal products, building materials, cosmetics, perfumes and incense, fibers, glazed ware, glass and its manufacture, materials used in the mummification process, and much more. 544pp. 6⅛ x 9¼. (Available in U.S. only.) 0-486-40446-3

RUSSIAN STORIES/RUSSKIE RASSKAZY: A Dual-Language Book, edited by Gleb Struve. Twelve tales by such masters as Chekhov, Tolstoy, Dostoevsky, Pushkin, others. Excellent word-for-word English translations on facing pages, plus teaching and study aids, Russian/English vocabulary, biographical/critical introductions, more. 416pp. 5⅜ x 8½. 0-486-26244-8

PHILADELPHIA THEN AND NOW: 60 Sites Photographed in the Past and Present, Kenneth Finkel and Susan Oyama. Rare photographs of City Hall, Logan Square, Independence Hall, Betsy Ross House, other landmarks juxtaposed with contemporary views. Captures changing face of historic city. Introduction. Captions. 128pp. 8¼ x 11. 0-486-25790-8

NORTH AMERICAN INDIAN LIFE: Customs and Traditions of 23 Tribes, Elsie Clews Parsons (ed.). 27 fictionalized essays by noted anthropologists examine religion, customs, government, additional facets of life among the Winnebago, Crow, Zuni, Eskimo, other tribes. 480pp. 6⅛ x 9¼. 0-486-27377-6

TECHNICAL MANUAL AND DICTIONARY OF CLASSICAL BALLET, Gail Grant. Defines, explains, comments on steps, movements, poses and concepts. 15-page pictorial section. Basic book for student, viewer. 127pp. 5⅜ x 8½. 0-486-21843-0

THE MALE AND FEMALE FIGURE IN MOTION: 60 Classic Photographic Sequences, Eadweard Muybridge. 60 true-action photographs of men and women walking, running, climbing, bending, turning, etc., reproduced from rare 19th-century masterpiece. vi + 121pp. 9 x 12. 0-486-24745-7

CATALOG OF DOVER BOOKS

ANIMALS: 1,419 Copyright-Free Illustrations of Mammals, Birds, Fish, Insects, etc., Jim Harter (ed.). Clear wood engravings present, in extremely lifelike poses, over 1,000 species of animals. One of the most extensive pictorial sourcebooks of its kind. Captions. Index. 284pp. 9 x 12. 0-486-23766-4

1001 QUESTIONS ANSWERED ABOUT THE SEASHORE, N. J. Berrill and Jacquelyn Berrill. Queries answered about dolphins, sea snails, sponges, starfish, fishes, shore birds, many others. Covers appearance, breeding, growth, feeding, much more. 305pp. 5¼ x 8¼. 0-486-23366-9

ATTRACTING BIRDS TO YOUR YARD, William J. Weber. Easy-to-follow guide offers advice on how to attract the greatest diversity of birds: birdhouses, feeders, water and waterers, much more. 96pp. 5³⁄₁₆ x 8¼. 0-486-28927-3

MEDICINAL AND OTHER USES OF NORTH AMERICAN PLANTS: A Historical Survey with Special Reference to the Eastern Indian Tribes, Charlotte Erichsen-Brown. Chronological historical citations document 500 years of usage of plants, trees, shrubs native to eastern Canada, northeastern U.S. Also complete identifying information. 343 illustrations. 544pp. 6½ x 9¼. 0-486-25951-X

STORYBOOK MAZES, Dave Phillips. 23 stories and mazes on two-page spreads: Wizard of Oz, Treasure Island, Robin Hood, etc. Solutions. 64pp. 8¼ x 11.
0-486-23628-5

AMERICAN NEGRO SONGS: 230 Folk Songs and Spirituals, Religious and Secular, John W. Work. This authoritative study traces the African influences of songs sung and played by black Americans at work, in church, and as entertainment. The author discusses the lyric significance of such songs as "Swing Low, Sweet Chariot," "John Henry," and others and offers the words and music for 230 songs. Bibliography. Index of Song Titles. 272pp. 6½ x 9¼. 0-486-40271-1

MOVIE-STAR PORTRAITS OF THE FORTIES, John Kobal (ed.). 163 glamor, studio photos of 106 stars of the 1940s: Rita Hayworth, Ava Gardner, Marlon Brando, Clark Gable, many more. 176pp. 8⅜ x 11¼. 0-486-23546-7

YEKL and THE IMPORTED BRIDEGROOM AND OTHER STORIES OF YIDDISH NEW YORK, Abraham Cahan. Film Hester Street based on Yekl (1896). Novel, other stories among first about Jewish immigrants on N.Y.'s East Side. 240pp. 5⅜ x 8½. 0-486-22427-9

SELECTED POEMS, Walt Whitman. Generous sampling from Leaves of Grass. Twenty-four poems include "I Hear America Singing," "Song of the Open Road," "I Sing the Body Electric," "When Lilacs Last in the Dooryard Bloom'd," "O Captain! My Captain!"–all reprinted from an authoritative edition. Lists of titles and first lines. 128pp. 5³⁄₁₆ x 8¼. 0-486-26878-0

SONGS OF EXPERIENCE: Facsimile Reproduction with 26 Plates in Full Color, William Blake. 26 full-color plates from a rare 1826 edition. Includes "The Tyger," "London," "Holy Thursday," and other poems. Printed text of poems. 48pp. 5¼ x 7.
0-486-24636-1

THE BEST TALES OF HOFFMANN, E. T. A. Hoffmann. 10 of Hoffmann's most important stories: "Nutcracker and the King of Mice," "The Golden Flowerpot," etc. 458pp. 5⅜ x 8½. 0-486-21793-0

THE BOOK OF TEA, Kakuzo Okakura. Minor classic of the Orient: entertaining, charming explanation, interpretation of traditional Japanese culture in terms of tea ceremony. 94pp. 5⅜ x 8½. 0-486-20070-1

FRENCH STORIES/CONTES FRANÇAIS: A Dual-Language Book, Wallace Fowlie. Ten stories by French masters, Voltaire to Camus: "Micromegas" by Voltaire; "The Atheist's Mass" by Balzac; "Minuet" by de Maupassant; "The Guest" by Camus, six more. Excellent English translations on facing pages. Also French-English vocabulary list, exercises, more. 352pp. 5⅜ x 8½. 0-486-26443-2

CHICAGO AT THE TURN OF THE CENTURY IN PHOTOGRAPHS: 122 Historic Views from the Collections of the Chicago Historical Society, Larry A. Viskochil. Rare large-format prints offer detailed views of City Hall, State Street, the Loop, Hull House, Union Station, many other landmarks, circa 1904-1913. Introduction. Captions. Maps. 144pp. 9⅜ x 12¼. 0-486-24656-6

OLD BROOKLYN IN EARLY PHOTOGRAPHS, 1865-1929, William Lee Younger. Luna Park, Gravesend race track, construction of Grand Army Plaza, moving of Hotel Brighton, etc. 157 previously unpublished photographs. 165pp. 8⅞ x 11¾. 0-486-23587-4

THE MYTHS OF THE NORTH AMERICAN INDIANS, Lewis Spence. Rich anthology of the myths and legends of the Algonquins, Iroquois, Pawnees and Sioux, prefaced by an extensive historical and ethnological commentary. 36 illustrations. 480pp. 5⅜ x 8½. 0-486-25967-6

AN ENCYCLOPEDIA OF BATTLES: Accounts of Over 1,560 Battles from 1479 B.C. to the Present, David Eggenberger. Essential details of every major battle in recorded history from the first battle of Megiddo in 1479 B.C. to Grenada in 1984. List of Battle Maps. New Appendix covering the years 1967-1984. Index. 99 illustrations. 544pp. 6½ x 9¼. 0-486-24913-1

SAILING ALONE AROUND THE WORLD, Captain Joshua Slocum. First man to sail around the world, alone, in small boat. One of great feats of seamanship told in delightful manner. 67 illustrations. 294pp. 5⅜ x 8½. 0-486-20326-3

ANARCHISM AND OTHER ESSAYS, Emma Goldman. Powerful, penetrating, prophetic essays on direct action, role of minorities, prison reform, puritan hypocrisy, violence, etc. 271pp. 5⅜ x 8½. 0-486-22484-8

MYTHS OF THE HINDUS AND BUDDHISTS, Ananda K. Coomaraswamy and Sister Nivedita. Great stories of the epics; deeds of Krishna, Shiva, taken from puranas, Vedas, folk tales; etc. 32 illustrations. 400pp. 5⅜ x 8½. 0-486-21759-0

MY BONDAGE AND MY FREEDOM, Frederick Douglass. Born a slave, Douglass became outspoken force in antislavery movement. The best of Douglass' autobiographies. Graphic description of slave life. 464pp. 5⅜ x 8½. 0-486-22457-0

FOLLOWING THE EQUATOR: A Journey Around the World, Mark Twain. Fascinating humorous account of 1897 voyage to Hawaii, Australia, India, New Zealand, etc. Ironic, bemused reports on peoples, customs, climate, flora and fauna, politics, much more. 197 illustrations. 720pp. 5⅜ x 8½. 0-486-26113-1

THE PEOPLE CALLED SHAKERS, Edward D. Andrews. Definitive study of Shakers: origins, beliefs, practices, dances, social organization, furniture and crafts, etc. 33 illustrations. 351pp. 5⅜ x 8½. 0-486-21081-2

THE MYTHS OF GREECE AND ROME, H. A. Guerber. A classic of mythology, generously illustrated, long prized for its simple, graphic, accurate retelling of the principal myths of Greece and Rome, and for its commentary on their origins and significance. With 64 illustrations by Michelangelo, Raphael, Titian, Rubens, Canova, Bernini and others. 480pp. 5⅜ x 8½. 0-486-27584-1

PSYCHOLOGY OF MUSIC, Carl E. Seashore. Classic work discusses music as a medium from psychological viewpoint. Clear treatment of physical acoustics, auditory apparatus, sound perception, development of musical skills, nature of musical feeling, host of other topics. 88 figures. 408pp. 5⅜ x 8½. 0-486-21851-1

LIFE IN ANCIENT EGYPT, Adolf Erman. Fullest, most thorough, detailed older account with much not in more recent books, domestic life, religion, magic, medicine, commerce, much more. Many illustrations reproduce tomb paintings, carvings, hieroglyphs, etc. 597pp. 5⅜ x 8½. 0-486-22632-8

SUNDIALS, Their Theory and Construction, Albert Waugh. Far and away the best, most thorough coverage of ideas, mathematics concerned, types, construction, adjusting anywhere. Simple, nontechnical treatment allows even children to build several of these dials. Over 100 illustrations. 230pp. 5⅜ x 8½. 0-486-22947-5

THEORETICAL HYDRODYNAMICS, L. M. Milne-Thomson. Classic exposition of the mathematical theory of fluid motion, applicable to both hydrodynamics and aerodynamics. Over 600 exercises. 768pp. 6⅛ x 9¼. 0-486-68970-0

OLD-TIME VIGNETTES IN FULL COLOR, Carol Belanger Grafton (ed.). Over 390 charming, often sentimental illustrations, selected from archives of Victorian graphics–pretty women posing, children playing, food, flowers, kittens and puppies, smiling cherubs, birds and butterflies, much more. All copyright-free. 48pp. 9¼ x 12¼. 0-486-27269-9

PERSPECTIVE FOR ARTISTS, Rex Vicat Cole. Depth, perspective of sky and sea, shadows, much more, not usually covered. 391 diagrams, 81 reproductions of drawings and paintings. 279pp. 5⅜ x 8½. 0-486-22487-2

DRAWING THE LIVING FIGURE, Joseph Sheppard. Innovative approach to artistic anatomy focuses on specifics of surface anatomy, rather than muscles and bones. Over 170 drawings of live models in front, back and side views, and in widely varying poses. Accompanying diagrams. 177 illustrations. Introduction. Index. 144pp. 8⅜ x11¼. 0-486-26723-7

GOTHIC AND OLD ENGLISH ALPHABETS: 100 Complete Fonts, Dan X. Solo. Add power, elegance to posters, signs, other graphics with 100 stunning copyright-free alphabets: Blackstone, Dolbey, Germania, 97 more–including many lower case, numerals, punctuation marks. 104pp. 8⅛ x 11. 0-486-24695-7

THE BOOK OF WOOD CARVING, Charles Marshall Sayers. Finest book for beginners discusses fundamentals and offers 34 designs. "Absolutely first rate . . . well thought out and well executed."–E. J. Tangerman. 118pp. 7¾ x 10⅝. 0-486-23654-4

ILLUSTRATED CATALOG OF CIVIL WAR MILITARY GOODS: Union Army Weapons, Insignia, Uniform Accessories, and Other Equipment, Schuyler, Hartley, and Graham. Rare, profusely illustrated 1846 catalog includes Union Army uniform and dress regulations, arms and ammunition, coats, insignia, flags, swords, rifles, etc. 226 illustrations. 160pp. 9 x 12. 0-486-24939-5

WOMEN'S FASHIONS OF THE EARLY 1900s: An Unabridged Republication of "New York Fashions, 1909," National Cloak & Suit Co. Rare catalog of mail-order fashions documents women's and children's clothing styles shortly after the turn of the century. Captions offer full descriptions, prices. Invaluable resource for fashion, costume historians. Approximately 725 illustrations. 128pp. 8⅜ x 11¼. 0-486-27276-1

HOW TO DO BEADWORK, Mary White. Fundamental book on craft from simple projects to five-bead chains and woven works. 106 illustrations. 142pp. 5⅜ x 8. 0-486-20697-1

THE 1912 AND 1915 GUSTAV STICKLEY FURNITURE CATALOGS, Gustav Stickley. With over 200 detailed illustrations and descriptions, these two catalogs are essential reading and reference materials and identification guides for Stickley furniture. Captions cite materials, dimensions and prices. 112pp. 6½ x 9¼. 0-486-26676-1

EARLY AMERICAN LOCOMOTIVES, John H. White, Jr. Finest locomotive engravings from early 19th century: historical (1804–74), main-line (after 1870), special, foreign, etc. 147 plates. 142pp. 11⅜ x 8¼. 0-486-22772-3

LITTLE BOOK OF EARLY AMERICAN CRAFTS AND TRADES, Peter Stockham (ed.). 1807 children's book explains crafts and trades: baker, hatter, cooper, potter, and many others. 23 copperplate illustrations. 140pp. 4⅝ x 6. 0-486-23336-7

VICTORIAN FASHIONS AND COSTUMES FROM HARPER'S BAZAR, 1867–1898, Stella Blum (ed.). Day costumes, evening wear, sports clothes, shoes, hats, other accessories in over 1,000 detailed engravings. 320pp. 9⅜ x 12¼. 0-486-22990-4

THE LONG ISLAND RAIL ROAD IN EARLY PHOTOGRAPHS, Ron Ziel. Over 220 rare photos, informative text document origin (1844) and development of rail service on Long Island. Vintage views of early trains, locomotives, stations, passengers, crews, much more. Captions. 8⅞ x 11¾. 0-486-26301-0

VOYAGE OF THE LIBERDADE, Joshua Slocum. Great 19th-century mariner's thrilling, first-hand account of the wreck of his ship off South America, the 35-foot boat he built from the wreckage, and its remarkable voyage home. 128pp. 5⅜ x 8½. 0-486-40022-0

TEN BOOKS ON ARCHITECTURE, Vitruvius. The most important book ever written on architecture. Early Roman aesthetics, technology, classical orders, site selection, all other aspects. Morgan translation. 331pp. 5⅜ x 8½. 0-486-20645-9

THE HUMAN FIGURE IN MOTION, Eadweard Muybridge. More than 4,500 stopped-action photos, in action series, showing undraped men, women, children jumping, lying down, throwing, sitting, wrestling, carrying, etc. 390pp. 7⅞ x 10⅝. 0-486-20204-6 Clothbd.

TREES OF THE EASTERN AND CENTRAL UNITED STATES AND CANADA, William M. Harlow. Best one-volume guide to 140 trees. Full descriptions, woodlore, range, etc. Over 600 illustrations. Handy size. 288pp. 4½ x 6⅜. 0-486-20395-6

GROWING AND USING HERBS AND SPICES, Milo Miloradovich. Versatile handbook provides all the information needed for cultivation and use of all the herbs and spices available in North America. 4 illustrations. Index. Glossary. 236pp. 5⅜ x 8½. 0-486-25058-X

BIG BOOK OF MAZES AND LABYRINTHS, Walter Shepherd. 50 mazes and labyrinths in all–classical, solid, ripple, and more–in one great volume. Perfect inexpensive puzzler for clever youngsters. Full solutions. 112pp. 8⅛ x 11. 0-486-22951-3

PIANO TUNING, J. Cree Fischer. Clearest, best book for beginner, amateur. Simple repairs, raising dropped notes, tuning by easy method of flattened fifths. No previous skills needed. 4 illustrations. 201pp. 5⅜ x 8½. 0-486-23267-0

HINTS TO SINGERS, Lillian Nordica. Selecting the right teacher, developing confidence, overcoming stage fright, and many other important skills receive thoughtful discussion in this indispensible guide, written by a world-famous diva of four decades' experience. 96pp. 5⅜ x 8½. 0-486-40094-8

THE COMPLETE NONSENSE OF EDWARD LEAR, Edward Lear. All nonsense limericks, zany alphabets, Owl and Pussycat, songs, nonsense botany, etc., illustrated by Lear. Total of 320pp. 5⅜ x 8½. (Available in U.S. only.) 0-486-20167-8

VICTORIAN PARLOUR POETRY: An Annotated Anthology, Michael R. Turner. 117 gems by Longfellow, Tennyson, Browning, many lesser-known poets. "The Village Blacksmith," "Curfew Must Not Ring Tonight," "Only a Baby Small," dozens more, often difficult to find elsewhere. Index of poets, titles, first lines. xxiii + 325pp. 5⅝ x 8¼. 0-486-27044-0

DUBLINERS, James Joyce. Fifteen stories offer vivid, tightly focused observations of the lives of Dublin's poorer classes. At least one, "The Dead," is considered a masterpiece. Reprinted complete and unabridged from standard edition. 160pp. 5$\frac{3}{16}$ x 8¼. 0-486-26870-5

GREAT WEIRD TALES: 14 Stories by Lovecraft, Blackwood, Machen and Others, S. T. Joshi (ed.). 14 spellbinding tales, including "The Sin Eater," by Fiona McLeod, "The Eye Above the Mantel," by Frank Belknap Long, as well as renowned works by R. H. Barlow, Lord Dunsany, Arthur Machen, W. C. Morrow and eight other masters of the genre. 256pp. 5⅜ x 8½. (Available in U.S. only.) 0-486-40436-6

THE BOOK OF THE SACRED MAGIC OF ABRAMELIN THE MAGE, translated by S. MacGregor Mathers. Medieval manuscript of ceremonial magic. Basic document in Aleister Crowley, Golden Dawn groups. 268pp. 5⅜ x 8½. 0-486-23211-5

THE BATTLES THAT CHANGED HISTORY, Fletcher Pratt. Eminent historian profiles 16 crucial conflicts, ancient to modern, that changed the course of civilization. 352pp. 5⅜ x 8½. 0-486-41129-X

NEW RUSSIAN-ENGLISH AND ENGLISH-RUSSIAN DICTIONARY, M. A. O'Brien. This is a remarkably handy Russian dictionary, containing a surprising amount of information, including over 70,000 entries. 366pp. 4½ x 6⅛. 0-486-20208-9

NEW YORK IN THE FORTIES, Andreas Feininger. 162 brilliant photographs by the well-known photographer, formerly with *Life* magazine. Commuters, shoppers, Times Square at night, much else from city at its peak. Captions by John von Hartz. 181pp. 9¼ x 10¾. 0-486-23585-8

INDIAN SIGN LANGUAGE, William Tomkins. Over 525 signs developed by Sioux and other tribes. Written instructions and diagrams. Also 290 pictographs. 111pp. 6⅛ x 9¼. 0-486-22029-X

ANATOMY: A Complete Guide for Artists, Joseph Sheppard. A master of figure drawing shows artists how to render human anatomy convincingly. Over 460 illustrations. 224pp. 8⅜ x 11¼. 0-486-27279-6

MEDIEVAL CALLIGRAPHY: Its History and Technique, Marc Drogin. Spirited history, comprehensive instruction manual covers 13 styles (ca. 4th century through 15th). Excellent photographs; directions for duplicating medieval techniques with modern tools. 224pp. 8⅜ x 11¼. 0-486-26142-5

DRIED FLOWERS: How to Prepare Them, Sarah Whitlock and Martha Rankin. Complete instructions on how to use silica gel, meal and borax, perlite aggregate, sand and borax, glycerine and water to create attractive permanent flower arrangements. 12 illustrations. 32pp. 5⅜ x 8½. 0-486-21802-3

EASY-TO-MAKE BIRD FEEDERS FOR WOODWORKERS, Scott D. Campbell. Detailed, simple-to-use guide for designing, constructing, caring for and using feeders. Text, illustrations for 12 classic and contemporary designs. 96pp. 5⅜ x 8½. 0-486-25847-5

THE COMPLETE BOOK OF BIRDHOUSE CONSTRUCTION FOR WOODWORKERS, Scott D. Campbell. Detailed instructions, illustrations, tables. Also data on bird habitat and instinct patterns. Bibliography. 3 tables. 63 illustrations in 15 figures. 48pp. 5¼ x 8½. 0-486-24407-5

SCOTTISH WONDER TALES FROM MYTH AND LEGEND, Donald A. Mackenzie. 16 lively tales tell of giants rumbling down mountainsides, of a magic wand that turns stone pillars into warriors, of gods and goddesses, evil hags, powerful forces and more. 240pp. 5⅜ x 8½. 0-486-29677-6

THE HISTORY OF UNDERCLOTHES, C. Willett Cunnington and Phyllis Cunnington. Fascinating, well-documented survey covering six centuries of English undergarments, enhanced with over 100 illustrations: 12th-century laced-up bodice, footed long drawers (1795), 19th-century bustles, 19th-century corsets for men, Victorian "bust improvers," much more. 272pp. 5⅝ x 8¼. 0-486-27124-2

ARTS AND CRAFTS FURNITURE: The Complete Brooks Catalog of 1912, Brooks Manufacturing Co. Photos and detailed descriptions of more than 150 now very collectible furniture designs from the Arts and Crafts movement depict davenports, settees, buffets, desks, tables, chairs, bedsteads, dressers and more, all built of solid, quarter-sawed oak. Invaluable for students and enthusiasts of antiques, Americana and the decorative arts. 80pp. 6½ x 9¼. 0-486-27471-3

WILBUR AND ORVILLE: A Biography of the Wright Brothers, Fred Howard. Definitive, crisply written study tells the full story of the brothers' lives and work. A vividly written biography, unparalleled in scope and color, that also captures the spirit of an extraordinary era. 560pp. 6⅛ x 9¼. 0-486-40297-5

THE ARTS OF THE SAILOR: Knotting, Splicing and Ropework, Hervey Garrett Smith. Indispensable shipboard reference covers tools, basic knots and useful hitches; handsewing and canvas work, more. Over 100 illustrations. Delightful reading for sea lovers. 256pp. 5⅜ x 8½. 0-486-26440-8

FRANK LLOYD WRIGHT'S FALLINGWATER: The House and Its History, Second, Revised Edition, Donald Hoffmann. A total revision–both in text and illustrations–of the standard document on Fallingwater, the boldest, most personal architectural statement of Wright's mature years, updated with valuable new material from the recently opened Frank Lloyd Wright Archives. "Fascinating"–*The New York Times*. 116 illustrations. 128pp. 9¼ x 10¾. 0-486-27430-6

PHOTOGRAPHIC SKETCHBOOK OF THE CIVIL WAR, Alexander Gardner. 100 photos taken on field during the Civil War. Famous shots of Manassas Harper's Ferry, Lincoln, Richmond, slave pens, etc. 244pp. 10⅝ x 8¼. 0-486-22731-6

FIVE ACRES AND INDEPENDENCE, Maurice G. Kains. Great back-to-the-land classic explains basics of self-sufficient farming. The one book to get. 95 illustrations. 397pp. 5⅜ x 8½. 0-486-20974-1

A MODERN HERBAL, Margaret Grieve. Much the fullest, most exact, most useful compilation of herbal material. Gigantic alphabetical encyclopedia, from aconite to zedoary, gives botanical information, medical properties, folklore, economic uses, much else. Indispensable to serious reader. 161 illustrations. 888pp. 6½ x 9¼. 2-vol. set. (Available in U.S. only.) Vol. I: 0-486-22798-7 Vol. II: 0-486-22799-5

HIDDEN TREASURE MAZE BOOK, Dave Phillips. Solve 34 challenging mazes accompanied by heroic tales of adventure. Evil dragons, people-eating plants, bloodthirsty giants, many more dangerous adversaries lurk at every twist and turn. 34 mazes, stories, solutions. 48pp. 8¼ x 11. 0-486-24566-7

LETTERS OF W. A. MOZART, Wolfgang A. Mozart. Remarkable letters show bawdy wit, humor, imagination, musical insights, contemporary musical world; includes some letters from Leopold Mozart. 276pp. 5⅜ x 8½. 0-486-22859-2

BASIC PRINCIPLES OF CLASSICAL BALLET, Agrippina Vaganova. Great Russian theoretician, teacher explains methods for teaching classical ballet. 118 illustrations. 175pp. 5⅜ x 8½. 0-486-22036-2

THE JUMPING FROG, Mark Twain. Revenge edition. The original story of The Celebrated Jumping Frog of Calaveras County, a hapless French translation, and Twain's hilarious "retranslation" from the French. 12 illustrations. 66pp. 5⅜ x 8½. 0-486-22686-7

BEST REMEMBERED POEMS, Martin Gardner (ed.). The 126 poems in this superb collection of 19th- and 20th-century British and American verse range from Shelley's "To a Skylark" to the impassioned "Renascence" of Edna St. Vincent Millay and to Edward Lear's whimsical "The Owl and the Pussycat." 224pp. 5⅜ x 8½. 0-486-27165-X

COMPLETE SONNETS, William Shakespeare. Over 150 exquisite poems deal with love, friendship, the tyranny of time, beauty's evanescence, death and other themes in language of remarkable power, precision and beauty. Glossary of archaic terms. 80pp. 5³⁄₁₆ x 8¼. 0-486-26686-9

HISTORIC HOMES OF THE AMERICAN PRESIDENTS, Second, Revised Edition, Irvin Haas. A traveler's guide to American Presidential homes, most open to the public, depicting and describing homes occupied by every American President from George Washington to George Bush. With visiting hours, admission charges, travel routes. 175 photographs. Index. 160pp. 8¼ x 11. 0-486-26751-2

THE WIT AND HUMOR OF OSCAR WILDE, Alvin Redman (ed.). More than 1,000 ripostes, paradoxes, wisecracks: Work is the curse of the drinking classes; I can resist everything except temptation; etc. 258pp. 5⅜ x 8½. 0-486-20602-5

SHAKESPEARE LEXICON AND QUOTATION DICTIONARY, Alexander Schmidt. Full definitions, locations, shades of meaning in every word in plays and poems. More than 50,000 exact quotations. 1,485pp. 6½ x 9¼. 2-vol. set. Vol. 1: 0-486-22726-X Vol. 2: 0-486-22727-8

SELECTED POEMS, Emily Dickinson. Over 100 best-known, best-loved poems by one of America's foremost poets, reprinted from authoritative early editions. No comparable edition at this price. Index of first lines. 64pp. 5³⁄₁₆ x 8¼. 0-486-26466-1

THE INSIDIOUS DR. FU-MANCHU, Sax Rohmer. The first of the popular mystery series introduces a pair of English detectives to their archnemesis, the diabolical Dr. Fu-Manchu. Flavorful atmosphere, fast-paced action, and colorful characters enliven this classic of the genre. 208pp. 5³⁄₁₆ x 8¼. 0-486-29898-1

THE MALLEUS MALEFICARUM OF KRAMER AND SPRENGER, translated by Montague Summers. Full text of most important witchhunter's "bible," used by both Catholics and Protestants. 278pp. 6⅝ x 10.   0-486-22802-9

SPANISH STORIES/CUENTOS ESPAÑOLES: A Dual-Language Book, Angel Flores (ed.). Unique format offers 13 great stories in Spanish by Cervantes, Borges, others. Faithful English translations on facing pages. 352pp. 5⅜ x 8½.

0-486-25399-6

GARDEN CITY, LONG ISLAND, IN EARLY PHOTOGRAPHS, 1869–1919, Mildred H. Smith. Handsome treasury of 118 vintage pictures, accompanied by carefully researched captions, document the Garden City Hotel fire (1899), the Vanderbilt Cup Race (1908), the first airmail flight departing from the Nassau Boulevard Aerodrome (1911), and much more. 96pp. 8⅞ x 11¾.   0-486-40669-5

OLD QUEENS, N.Y., IN EARLY PHOTOGRAPHS, Vincent F. Seyfried and William Asadorian. Over 160 rare photographs of Maspeth, Jamaica, Jackson Heights, and other areas. Vintage views of DeWitt Clinton mansion, 1939 World's Fair and more. Captions. 192pp. 8⅞ x 11.   0-486-26358-4

CAPTURED BY THE INDIANS: 15 Firsthand Accounts, 1750-1870, Frederick Drimmer. Astounding true historical accounts of grisly torture, bloody conflicts, relentless pursuits, miraculous escapes and more, by people who lived to tell the tale. 384pp. 5⅜ x 8½.   0-486-24901-8

THE WORLD'S GREAT SPEECHES (Fourth Enlarged Edition), Lewis Copeland, Lawrence W. Lamm, and Stephen J. McKenna. Nearly 300 speeches provide public speakers with a wealth of updated quotes and inspiration–from Pericles' funeral oration and William Jennings Bryan's "Cross of Gold Speech" to Malcolm X's powerful words on the Black Revolution and Earl of Spenser's tribute to his sister, Diana, Princess of Wales. 944pp. 5⅜ x 8⅜.   0-486-40903-1

THE BOOK OF THE SWORD, Sir Richard F. Burton. Great Victorian scholar/adventurer's eloquent, erudite history of the "queen of weapons"–from prehistory to early Roman Empire. Evolution and development of early swords, variations (sabre, broadsword, cutlass, scimitar, etc.), much more. 336pp. 6⅛ x 9¼.

0-486-25434-8

AUTOBIOGRAPHY: The Story of My Experiments with Truth, Mohandas K. Gandhi. Boyhood, legal studies, purification, the growth of the Satyagraha (nonviolent protest) movement. Critical, inspiring work of the man responsible for the freedom of India. 480pp. 5⅜ x 8½. (Available in U.S. only.)   0-486-24593-4

CELTIC MYTHS AND LEGENDS, T. W. Rolleston. Masterful retelling of Irish and Welsh stories and tales. Cuchulain, King Arthur, Deirdre, the Grail, many more. First paperback edition. 58 full-page illustrations. 512pp. 5⅜ x 8½.   0-486-26507-2

THE PRINCIPLES OF PSYCHOLOGY, William James. Famous long course complete, unabridged. Stream of thought, time perception, memory, experimental methods; great work decades ahead of its time. 94 figures. 1,391pp. 5⅜ x 8½. 2-vol. set.
Vol. I: 0-486-20381-6   Vol. II: 0-486-20382-4

THE WORLD AS WILL AND REPRESENTATION, Arthur Schopenhauer. Definitive English translation of Schopenhauer's life work, correcting more than 1,000 errors, omissions in earlier translations. Translated by E. F. J. Payne. Total of 1,269pp. 5⅜ x 8½. 2-vol. set.   Vol. 1: 0-486-21761-2   Vol. 2: 0-486-21762-0

CATALOG OF DOVER BOOKS

MAGIC AND MYSTERY IN TIBET, Madame Alexandra David-Neel. Experiences among lamas, magicians, sages, sorcerers, Bonpa wizards. A true psychic discovery. 32 illustrations. 321pp. 5⅜ x 8½. (Available in U.S. only.)   0-486-22682-4

THE EGYPTIAN BOOK OF THE DEAD, E. A. Wallis Budge. Complete reproduction of Ani's papyrus, finest ever found. Full hieroglyphic text, interlinear transliteration, word-for-word translation, smooth translation. 533pp. 6½ x 9¼.
0-486-21866-X

HISTORIC COSTUME IN PICTURES, Braun & Schneider. Over 1,450 costumed figures in clearly detailed engravings–from dawn of civilization to end of 19th century. Captions. Many folk costumes. 256pp. 8⅜ x 11¼.   0-486-23150-X

MATHEMATICS FOR THE NONMATHEMATICIAN, Morris Kline. Detailed, college-level treatment of mathematics in cultural and historical context, with numerous exercises. Recommended Reading Lists. Tables. Numerous figures. 641pp. 5⅜ x 8½.
0-486-24823-2

PROBABILISTIC METHODS IN THE THEORY OF STRUCTURES, Isaac Elishakoff. Well-written introduction covers the elements of the theory of probability from two or more random variables, the reliability of such multivariable structures, the theory of random function, Monte Carlo methods of treating problems incapable of exact solution, and more. Examples. 502pp. 5⅜ x 8½.   0-486-40691-1

THE RIME OF THE ANCIENT MARINER, Gustave Doré, S. T. Coleridge. Doré's finest work; 34 plates capture moods, subtleties of poem. Flawless full-size reproductions printed on facing pages with authoritative text of poem. "Beautiful. Simply beautiful."–*Publisher's Weekly.* 77pp. 9¼ x 12.   0-486-22305-1

SCULPTURE: Principles and Practice, Louis Slobodkin. Step-by-step approach to clay, plaster, metals, stone; classical and modern. 253 drawings, photos. 255pp. 8⅛ x 11.
0-486-22960-2

THE INFLUENCE OF SEA POWER UPON HISTORY, 1660–1783, A. T. Mahan. Influential classic of naval history and tactics still used as text in war colleges. First paperback edition. 4 maps. 24 battle plans. 640pp. 5⅜ x 8½.   0-486-25509-3

THE STORY OF THE TITANIC AS TOLD BY ITS SURVIVORS, Jack Winocour (ed.). What it was really like. Panic, despair, shocking inefficiency, and a little heroism. More thrilling than any fictional account. 26 illustrations. 320pp. 5⅜ x 8½.
0-486-20610-6

ONE TWO THREE . . . INFINITY: Facts and Speculations of Science, George Gamow. Great physicist's fascinating, readable overview of contemporary science: number theory, relativity, fourth dimension, entropy, genes, atomic structure, much more. 128 illustrations. Index. 352pp. 5⅜ x 8½.   0-486-25664-2

DALÍ ON MODERN ART: The Cuckolds of Antiquated Modern Art, Salvador Dalí. Influential painter skewers modern art and its practitioners. Outrageous evaluations of Picasso, Cézanne, Turner, more. 15 renderings of paintings discussed. 44 calligraphic decorations by Dalí. 96pp. 5⅜ x 8½. (Available in U.S. only.)   0-486-29220-7

ANTIQUE PLAYING CARDS: A Pictorial History, Henry René D'Allemagne. Over 900 elaborate, decorative images from rare playing cards (14th–20th centuries): Bacchus, death, dancing dogs, hunting scenes, royal coats of arms, players cheating, much more. 96pp. 9¼ x 12¼.   0-486-29265-7

MAKING FURNITURE MASTERPIECES: 30 Projects with Measured Drawings, Franklin H. Gottshall. Step-by-step instructions, illustrations for constructing handsome, useful pieces, among them a Sheraton desk, Chippendale chair, Spanish desk, Queen Anne table and a William and Mary dressing mirror. 224pp. 8⅛ x 11¼. 0-486-29338-6

NORTH AMERICAN INDIAN DESIGNS FOR ARTISTS AND CRAFTSPEOPLE, Eva Wilson. Over 360 authentic copyright-free designs adapted from Navajo blankets, Hopi pottery, Sioux buffalo hides, more. Geometrics, symbolic figures, plant and animal motifs, etc. 128pp. 8⅜ x 11. (Not for sale in the United Kingdom.) 0-486-25341-4

THE FOSSIL BOOK: A Record of Prehistoric Life, Patricia V. Rich et al. Profusely illustrated definitive guide covers everything from single-celled organisms and dinosaurs to birds and mammals and the interplay between climate and man. Over 1,500 illustrations. 760pp. 7½ x 10⅛. 0-486-29371-8

VICTORIAN ARCHITECTURAL DETAILS: Designs for Over 700 Stairs, Mantels, Doors, Windows, Cornices, Porches, and Other Decorative Elements, A. J. Bicknell & Company. Everything from dormer windows and piazzas to balconies and gable ornaments. Also includes elevations and floor plans for handsome, private residences and commercial structures. 80pp. 9⅜ x 12¼. 0-486-44015-X

WESTERN ISLAMIC ARCHITECTURE: A Concise Introduction, John D. Hoag. Profusely illustrated critical appraisal compares and contrasts Islamic mosques and palaces–from Spain and Egypt to other areas in the Middle East. 139 illustrations. 128pp. 6 x 9. 0-486-43760-4

CHINESE ARCHITECTURE: A Pictorial History, Liang Ssu-ch'eng. More than 240 rare photographs and drawings depict temples, pagodas, tombs, bridges, and imperial palaces comprising much of China's architectural heritage. 152 halftones, 94 diagrams. 232pp. 10¾ x 9⅞. 0-486-43999-2

THE RENAISSANCE: Studies in Art and Poetry, Walter Pater. One of the most talked-about books of the 19th century, *The Renaissance* combines scholarship and philosophy in an innovative work of cultural criticism that examines the achievements of Botticelli, Leonardo, Michelangelo, and other artists. "The holy writ of beauty."–Oscar Wilde. 160pp. 5⅜ x 8½. 0-486-44025-7

A TREATISE ON PAINTING, Leonardo da Vinci. The great Renaissance artist's practical advice on drawing and painting techniques covers anatomy, perspective, composition, light and shadow, and color. A classic of art instruction, it features 48 drawings by Nicholas Poussin and Leon Battista Alberti. 192pp. 5⅜ x 8½. 0-486-44155-5

THE MIND OF LEONARDO DA VINCI, Edward McCurdy. More than just a biography, this classic study by a distinguished historian draws upon Leonardo's extensive writings to offer numerous demonstrations of the Renaissance master's achievements, not only in sculpture and painting, but also in music, engineering, and even experimental aviation. 384pp. 5⅜ x 8½. 0-486-44142-3

WASHINGTON IRVING'S RIP VAN WINKLE, Illustrated by Arthur Rackham. Lovely prints that established artist as a leading illustrator of the time and forever etched into the popular imagination a classic of Catskill lore. 51 full-color plates. 80pp. 8⅜ x 11. 0-486-44242-X

HENSCHE ON PAINTING, John W. Robichaux. Basic painting philosophy and methodology of a great teacher, as expounded in his famous classes and workshops on Cape Cod. 7 illustrations in color on covers. 80pp. 5⅜ x 8½. 0-486-43728-0

LIGHT AND SHADE: A Classic Approach to Three-Dimensional Drawing, Mrs. Mary P. Merrifield. Handy reference clearly demonstrates principles of light and shade by revealing effects of common daylight, sunshine, and candle or artificial light on geometrical solids. 13 plates. 64pp. 5⅜ x 8½. 0-486-44143-1

ASTROLOGY AND ASTRONOMY: A Pictorial Archive of Signs and Symbols, Ernst and Johanna Lehner. Treasure trove of stories, lore, and myth, accompanied by more than 300 rare illustrations of planets, the Milky Way, signs of the zodiac, comets, meteors, and other astronomical phenomena. 192pp. 8⅜ x 11. 0-486-43981-X

JEWELRY MAKING: Techniques for Metal, Tim McCreight. Easy-to-follow instructions and carefully executed illustrations describe tools and techniques, use of gems and enamels, wire inlay, casting, and other topics. 72 line illustrations and diagrams. 176pp. 8¼ x 10⅞. 0-486-44043-5

MAKING BIRDHOUSES: Easy and Advanced Projects, Gladstone Califf. Easy-to-follow instructions include diagrams for everything from a one-room house for bluebirds to a forty-two-room structure for purple martins. 56 plates; 4 figures. 80pp. 8¾ x 6⅜. 0-486-44183-0

LITTLE BOOK OF LOG CABINS: How to Build and Furnish Them, William S. Wicks. Handy how-to manual, with instructions and illustrations for building cabins in the Adirondack style, fireplaces, stairways, furniture, beamed ceilings, and more. 102 line drawings. 96pp. 8¾ x 6⅜. 0-486-44259-4

THE SEASONS OF AMERICA PAST, Eric Sloane. From "sugaring time" and strawberry picking to Indian summer and fall harvest, a whole year's activities described in charming prose and enhanced with 79 of the author's own illustrations. 160pp. 8¼ x 11. 0-486-44220-9

THE METROPOLIS OF TOMORROW, Hugh Ferriss. Generous, prophetic vision of the metropolis of the future, as perceived in 1929. Powerful illustrations of towering structures, wide avenues, and rooftop parks–all features in many of today's modern cities. 59 illustrations. 144pp. 8¼ x 11. 0-486-43727-2

THE PATH TO ROME, Hilaire Belloc. This 1902 memoir abounds in lively vignettes from a vanished time, recounting a pilgrimage on foot across the Alps and Apennines in order to "see all Europe which the Christian Faith has saved." 77 of the author's original line drawings complement his sparkling prose. 272pp. 5⅜ x 8½. 0-486-44001-X

THE HISTORY OF RASSELAS: Prince of Abissinia, Samuel Johnson. Distinguished English writer attacks eighteenth-century optimism and man's unrealistic estimates of what life has to offer. 112pp. 5⅜ x 8½. 0-486-44094-X

A VOYAGE TO ARCTURUS, David Lindsay. A brilliant flight of pure fancy, where wild creatures crowd the fantastic landscape and demented torturers dominate victims with their bizarre mental powers. 272pp. 5⅜ x 8½. 0-486-44198-9

Paperbound unless otherwise indicated. Available at your book dealer, online at **www.doverpublications.com**, or by writing to Dept. GI, Dover Publications, Inc., 31 East 2nd Street, Mineola, NY 11501. For current price information or for free catalogs (please indicate field of interest), write to Dover Publications or log on to **www.doverpublications.com** and see every Dover book in print. Dover publishes more than 500 books each year on science, elementary and advanced mathematics, biology, music, art, literary history, social sciences, and other areas.

ETON COUNTY LIBRARY
ACKSON, WYOMING